T0327364

Advanced Ceramic Coatings and Interfaces III

Advanced Ceramic Coatings and Interfaces III

A Collection of Papers Presented at the 32nd International Conference on Advanced Ceramics and Composites
January 27–February 1, 2008
Daytona Beach, Florida

Editors
Hua-Tay Lin
Dongming Zhu

Volume Editors
Tatsuki Ohji
Andrew Wereszczak

A John Wiley & Sons, Inc., Publication

Published by John Wiley & Sons, Inc., Hoboken, New Jersey.
Published simultaneously in Canada.

For general information on our other products and services or for technical support, please contact our Customer Care Department within the United States at (800) 762-2974, outside the United States at (317) 572-3993 or fax (317) 572-4002.

Wiley also publishes its books in a variety of electronic formats. Some content that appears in print may not be available in electronic format. For information about Wiley products, visit our web site at www.wiley.com.

Library of Congress Cataloging-in-Publication Data is available.

ISBN 978-0-470-34495-8

10 9 8 7 6 5 4 3 2 1

Contents

NANOSTRUCTURED COATINGS

THERMAL BARRIER COATING PROCESSING, DEVELOPMENT AND MODELING

Preface

The symposium on Advanced Ceramic Coatings for Structural, Environmental and Functional Applications was held during The American Ceramic Society's 32nd International Conference on Advanced Ceramics and Composites in Daytona Beach, Florida, January 28 to Feb 1, 2008. A total of 69 papers, including 8 invited talks, were presented at the symposium, covering broad ceramic coating and interface topic areas and emphasizing the latest advancement in coating processing, characterization, development, and applications.

The present volume contains fourteen contributed papers from the symposium, with topics including damping and erosion coatings, wear and tribological coatings, nanostructured coatings, and thermal barrier coating processing, development, modeling and life prediction, which highlights the state-of-the-art ceramic coatings technologies for various critical engineering applications.

We are greatly indebt to the members of the symposium organizing committee, including Drs.Yutaka Kagawa, Anette Karlsson, Irene Spitsberg, Dileep Singh, Yong-Ho Sohn, Xingbo Liu, Uwe Schulz, Robert Vaßen, and Jennifer Sample, for their assistance in developing and organizing this vibrant and cutting-edge symposium. We also would like to express our sincere thanks to manuscript authors and reviewers, all the symposium participants and session chairs for their contributions to a successful meeting. Finally, we are also grateful to the staff of The American Ceramic Society for their time and effort in ensuring an enjoyable conference and the high-quality publication of the proceeding volume.

H. T. Lin
Oak Ridge National Laboratory

Dongming Zhu
NASA Glenn Research Center

Introduction

Organized by the Engineering Ceramics Division (ECD) in conjunction with the Basic Science Division (BSD) of The American Ceramic Society (ACerS), the 32nd International Conference on Advanced Ceramics and Composites (ICACC) was held on January 27 to February 1, 2008, in Daytona Beach, Florida. 2008 was the second year that the meeting venue changed from Cocoa Beach, where ICACC was originated in January 1977 and was fostered to establish a meeting that is today the most preeminent international conference on advanced ceramics and composites

The 32nd ICACC hosted 1,247 attendees from 40 countries and 724 presentations on topics ranging from ceramic nanomaterials to structural reliability of ceramic components, demonstrating the linkage between materials science developments at the atomic level and macro level structural applications. The conference was organized into the following symposia and focused sessions:

Symposium 1	Mechanical Behavior and Structural Design of Monolithic and Composite Ceramics
Symposium 2	Advanced Ceramic Coatings for Structural, Environmental, and Functional Applications
Symposium 3	5th International Symposium on Solid Oxide Fuel Cells (SOFC): Materials, Science, and Technology
Symposium 4	Ceramic Armor
Symposium 5	Next Generation Bioceramics
Symposium 6	2nd International Symposium on Thermoelectric Materials for Power Conversion Applications
Symposium 7	2nd International Symposium on Nanostructured Materials and Nanotechnology: Development and Applications
Symposium 8	Advanced Processing & Manufacturing Technologies for Structural & Multifunctional Materials and Systems (APMT): An International Symposium in Honor of Prof. Yoshinari Miyamoto
Symposium 9	Porous Ceramics: Novel Developments and Applications

Symposium 10	Basic Science of Multifunctional Ceramics
Symposium 11	Science of Ceramic Interfaces: An International Symposium Memorializing Dr. Rowland M. Cannon
Focused Session 1	Geopolymers
Focused Session 2	Materials for Solid State Lighting

Peer reviewed papers were divided into nine issues of the 2008 Ceramic Engineering & Science Proceedings (CESP); Volume 29, Issues 2-10, as outlined below:

- Mechanical Properties and Processing of Ceramic Binary, Ternary and Composite Systems, Vol. 29, Is 2 (includes papers from symposium 1)
- Corrosion, Wear, Fatigue, and Reliability of Ceramics, Vol. 29, Is 3 (includes papers from symposium 1)
- Advanced Ceramic Coatings and Interfaces III, Vol. 29, Is 4 (includes papers from symposium 2)
- Advances in Solid Oxide Fuel Cells IV, Vol. 29, Is 5 (includes papers from symposium 3)
- Advances in Ceramic Armor IV, Vol. 29, Is 6 (includes papers from symposium 4)
- Advances in Bioceramics and Porous Ceramics, Vol. 29, Is 7 (includes papers from symposia 5 and 9)
- Nanostructured Materials and Nanotechnology II, Vol. 29, Is 8 (includes papers from symposium 7)
- Advanced Processing and Manufacturing Technologies for Structural and Multifunctional Materials II, Vol. 29, Is 9 (includes papers from symposium 8)
- Developments in Strategic Materials, Vol. 29, Is 10 (includes papers from symposia 6, 10, and 11, and focused sessions 1 and 2)

The organization of the Daytona Beach meeting and the publication of these proceedings were possible thanks to the professional staff of ACerS and the tireless dedication of many ECD and BSD members. We would especially like to express our sincere thanks to the symposia organizers, session chairs, presenters and conference attendees, for their efforts and enthusiastic participation in the vibrant and cutting-edge conference.

ACerS and the ECD invite you to attend the 33rd International Conference on Advanced Ceramics and Composites (http://www.ceramics.org/daytona2009) January 18–23, 2009 in Daytona Beach, Florida.

TATSUKI OHJI and ANDREW A. WERESZCZAK, Volume Editors
July 2008

Damping and Erosion Coatings

COATINGS FOR ENHANCED PASSIVE DAMPING

Peter J. Torvik
Professor Emeritus, Air Force Institute of Technology
Xenia OH, USA

ABSTRACT

The amplitude of vibration in a structure undergoing resonant vibration is governed by the total damping of the system. As the inherent damping of materials suitable for use in the fabrication of structures and machine components is often quite low, increasing the total system damping by including a dissipative material or mechanism can often provide significant reductions in the peak values of response (stress, strain, and displacement), enabling more efficient designs and enhanced performance. Available methodologies include active dampers and such passive techniques as friction and impact dampers and constrained layer treatments. It has also been found that metals and ceramics applied as free-layer hard coatings by plasma spray or electron beam physical vapor deposition add significant damping to vibrating members. In order to incorporate the influence of a damping coating in a prediction of system response during a preliminary design, it is essential that properties of the coating be known. The relevant damping characteristic is a measure of the energy dissipated by a homogeneous unit volume of material undergoing a completely reversed cycle of oscillation. A useful metric for this is the loss modulus. As all of these materials are inherently non-linear, as evidenced by amplitude-dependent measures of damping, determinations of properties must be made at the levels of strain appropriate to the application. Methods for determining the damping properties of materials are discussed, and comparisons are made of the damping of various classes of materials with those of ceramic coatings deposited by plasma spray or electron beam physical vapor deposition.

I. THE NEED FOR DAMPING

Although the assumption of a perfectly elastic material is very convenient for use in the analysis of structures, and adequate in most cases, no structural materials are truly elastic. A system given an initial perturbation will eventually come to rest unless the dissipation is offset by the addition of energy. Cyclic motion of a structure can be sustained at constant amplitude only if the energy lost through dissipation is offset by work done on the system. Energy dissipation, or damping, can be advantageous to the performance of a system as it governs the maximum amplitude achieved under resonance and the rate at which a perturbed system progresses to a satisfactorily quiescent state. In addition to lowering the probability of failure due to fatigue, a reduction of amplitude can have other benefits such as reducing a visible vibration, reducing the sound transmitted from a valve cover, or reducing the signature of a vibrating submarine propeller. Damping, however, can also be disadvantageous, contributing as it does to such unwanted phenomena as shaft whirl, instrument hysteresis, and temperature increases due to self-heating.

The term damping (not dampening) refers to the dissipation of energy in a material or structure under cyclic stress or strain through a process of converting mechanical energy (strain and kinetic) to heat. When such dissipation occurs locally within the material, the process is referred to as material damping, taken as inclusive of the dissipative mechanisms variously referred to as mechanical hysteresis, anelesticity, or internal friction.

The distinction between material damping and system damping should be observed. Material damping, the inherent ability of a substance to dissipate energy under cyclic stress or strain, is a material property and can be expressed in absolute units of energy dissipated per unit volume per cycle or as a dimensionless measure formed from a ratio of the energy dissipated to the energy stored. System damping, on the other hand, is a measure of the influence of the total dissipation of all the components of the system on the overall response of a structure or structural component.

At the present time, there is a particularly high level of interest in reducing the amplitude of vibration of the blades in gas turbine engines. As each blade in a rotating component passes a blade (stator) in the static component, a pressure pulse is generated. As the blades rotate at a high cyclic rate (~10,000 RPM) and there may well be several dozen stators, excitation rates of several thousand impulses per second are common. As a response at around 3K Hz gives rise to one million cycles of vibration in only 10 engine hours, and as the maximum velocity of structural motion is proportional to maximum strain, it is evident that excitation at a resonant frequency can lead to failure of the blade by fatigue. This phenomenon is known as high cycle fatigue (HCF) and is to be distinguished from the blade damage known as low cycle fatigue (LCF) that results from the much less frequently occurring perturbations due to major changes in engine RPM and temperature.

As it is the amount of damping present in the system that governs the magnitude of the response at resonance, it would be highly advantageous to be able to apply a passive damping treatment to rotating turbine blades. However, a successful treatment must survive not only the high vibratory stresses for the service life of the engine but also the environmental challenges of high temperature, erosion, corrosion, foreign object damage, and centrifugal forces due to the high rotation rates. Further, a successful treatment must not degrade the performance of the engine by adding excessive weight, by detrimental changes in the shapes of aerodynamic surfaces, or by inducing cracks in the blade material.

A number of concepts for the reduction of vibratory response are being explored. These include the use of friction dampers, impact dampers, energy absorbers, and the inclusion of dissipative materials in constrained layer damping treatments. Of a special interest is the use of free layer coatings in the form of thin layers of high damping metals and ceramics applied through plasma spray and electron beam physical vapor deposition. The use of such ceramics as alumina, magnesium aluminate spinel, and yttria stabilized zirconia appear to have received the greatest attention.

II. DAMPING AS A MATERIAL PROPERTY

A. Measures of Material Damping

The most fundamental measure of the dissipative capability of a material is the specific damping energy or unit damping, defined as the energy dissipated in a unit volume of material at homogeneous strain and temperature while undergoing a fully reversed cycle of cyclic stress or strain. The specific damping energy, D, has dimension of energy per unit volume, per cycle, and is, in general, a function of the amplitude and history of stress or strain, temperature, and frequency. For some materials, the unit damping is also dependent on the mean (static) stress or is influenced by magnetic fields. The unit damping is customarily given in terms of the amplitude of a uniaxial tensile or shear stress or strain. Multiaxial states of stress are characterized by an equivalent uniaxial stress[1].

Material damping may be categorized as being linear or non-linear. In the first class, the energy dissipated per cycle is dependent on the square of the amplitude of cyclic stress or strain. As the strain energy density is also normally proportional to the square of amplitude, the ratio of dissipated and stored energies, as well as other dimensionless measures of damping, are then independent of amplitude. Materials displaying these attributes are said to display linear damping. In the second class of materials, the energy dissipated per cycle varies as amplitude of cyclic stress to some power other than two. If the strain energy density varies as, or nearly as, the square of amplitude, the ratio of dissipated to stored energy is then a function of the amplitude of stress or strain. Such materials are said to display nonlinear damping.

While some important mechanisms of damping, such as viscoelastic and thermoelastic, are essentially linear, many others are not. In the case of structural materials for which the predominant damping mechanism is plastic deformation on a scale which leaves the material macroscopically linear, the specific damping energy has been found[1] to depend on the amplitude of cyclic stress as

$$D = J\sigma_d^n \tag{1}$$

with n ≈ 2.4 for cyclic stresses below about 70% of the endurance limit. For metals, the parameters J and n are generally independent of frequency, but dependent on temperature. At higher stress, the same functional form may be applied, but the damping typically increases more rapidly with stress. The parameter n is then much greater and may increase or decrease with the number of cycles.

While rooted in the concept of a linear viscoelastic material, the concept of the complex modulus may be adapted to characterize the dissipation of other materials undergoing cyclic loading. In the case of a nonlinear material, we may define amplitude-dependent effective values of a storage and a loss modulus, E_1 and E_2, by

$$E_1(\omega,T,\varepsilon_d) \equiv \frac{2U(\omega,T,\varepsilon_d)}{\varepsilon_d^2} \quad \text{and} \quad E_2(\omega,T,\varepsilon_d) \equiv \frac{D(\omega,T,\varepsilon_d)}{\pi\varepsilon_d^2} \tag{2a, b}$$

where U is the stored (strain) energy in the unit volume, D is the specific damping energy, and ε_d is the amplitude of cyclic strain. For structural materials, the values of storage and loss modulus are typically independent of frequency, but vary with amplitude and temperature. In the case of viscoelastic materials, the moduli are typically independent of amplitude, but vary strongly with both frequency and temperature. A material loss factor may also be defined as the ratio of energy dissipated in the unit volume per radian of oscillation to the peak energy stored.

$$\eta \equiv \frac{D}{2\pi U} = \frac{1}{2\pi}\frac{\pi E_2(\omega,T,\varepsilon_d)}{E_1(\omega,T,\varepsilon_d)/2} = \frac{E_2(\omega,T,\varepsilon_d)}{E_1(\omega,T,\varepsilon_d)} \tag{3}$$

If either or both of the components of the modulus are dependent on amplitude, then the material loss factor is also dependent on amplitude. Note that the use of the loss factor, defined in terms of energy dissipated per radian, is to be preferred over the sometimes-used specific damping capacity or damping index, computed from the energy dissipated per cycle by $\Psi = D/U$. This is because the unit of the radian is more truly a dimensionless quantity than is the cycle.

The complex modulus is defined as the ratio of the Fourier transform of stress to that of strain,

$$E^* = \sigma^*(\omega)/\varepsilon^*(\omega) = E_1 + jE_2 = E_1(1 + j\eta) \tag{4}$$

and is particularly convenient for analyzing the response of time-dependent materials to sinusoidal forcing functions. The storage modulus, E_1, is the customary Young's or elastic modulus; the loss modulus, E_2, is the product of the storage modulus and the loss factor. The use of a complex quantity to describe a real, physical material is troubling to some. The proper interpretation is not that the material behaves in a mathematically complex manner, but rather that the presence of dissipation is associated with an out-of-phase relationship between stress and strain. The loss factor quantifies the degree to which they are out of phase. The angle by which the strain lags the stress is given by $\tan\phi = E_2(\omega,T,\varepsilon_d)/E_1(\omega,T,\varepsilon_d)$ and is referred to as the loss tangent.

B. High Damping Materials

1. HIDAMETS

The specific damping energy of most structural materials falls in a fairly narrow range, particularly at stresses of design interest, i.e. well below the endurance limit. The mean curve for a wide range[1] of structural metals may be represented by $D = 14(\sigma_d / \sigma_E)^{2.4}$ Joule/m^3 for applied stresses, σ_d, up to about 70% of the endurance limit, σ_E. There are, however, a few exceptions, as such magnetoelastic materials as 403 alloy and the alloys of nickel and cobalt; alloys of manganese and copper; and the shape memory (SM) alloys. The high damping of magnetoelastic materials arises because the application of stress induces rotations of the magnetic domains similar to that produced by the application of a magnetic field, a process that dissipates energy. The high damping of Mn-Cu and the shape memory alloys is associated with transitions from face centered cubic to face centered tetragonal structures. Attempts to exploit these mechanisms so as to obtain high inherent material damping while meeting the other requirements of a material for primary structures have not, to date, been successful. These, and other high damping metals, may yet find application, perhaps as coatings.

2. Viscoelastic Materials

Another class of materials finding extensive application in structural additions for enhanced damping includes the elastomers and polymers. These are characterized by stress-strain relationships incorporating rate dependence. In the classical form, the constitutive relationship is taken to be a linear operator on stress proportional to a linear operator on strain. While this provides an adequate description of observed material properties, it has been found that many terms (necessitating many experimentally determined parameters) are required. More recently, it has been shown[2] that modifying the form by replacing the integer derivatives in the linear operators with fractional derivatives enables an excellent characterization over as many as 11 orders of magnitude in frequency. A characteristic of viscoelastic materials is that, in addition to having a strong dependence on frequency, they display a strong dependence on temperature. Fortunately, these are related. The concept of the reduced frequency has been accepted as a means of jointly describing both effects. A complete discussion is available elsewhere.[3]

As the ability to dissipate energy is dependent on the loss modulus, Eq. 2b, the loss modulus is an appropriate metric for identifying high damping materials. The solid line of Figure 1 represents a constant value of loss modulus (product of loss factor and storage modulus), with 1 GPa used as a reference value. Structural metals are seen to generally fall below this line; materials considered for treatments to enhance damping, such as the high damping alloys and magnetoelastic materials, fall on or above this line.

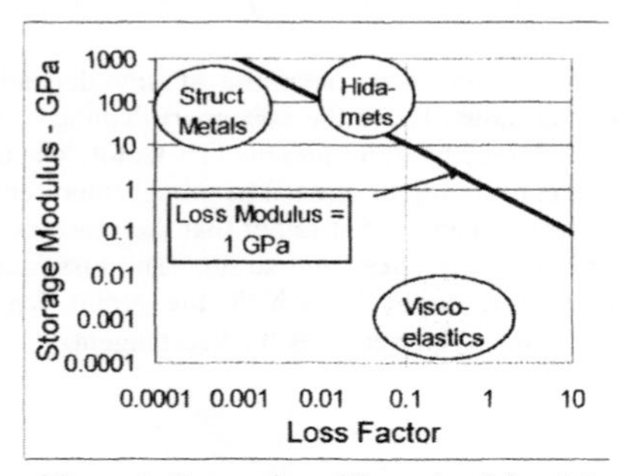

Figure 1. Categories of Damping Materials

Dissipation depends not only on the loss modulus, but also on the square of strain. As many of the viscoelastic materials can safely withstand repetitive strain on the order of unity, they most truly earn their reputation as high damping materials when used in a damping treatment with configuration designed so as induce strains in the viscoelastic material that are 2-3 orders of magnitude more than that of the base structure.

III. DAMPING AS A SYSTEM PROPERTY

The damping of systems containing one or more dissipating elements is of interest for two reasons. In order to incorporate damping into predictions of the system performance during the preliminary design process, a mathematical model of the system, including damping, is necessary. Secondly, as it is typically very difficult to perform materials testing on samples of a homogeneous material in a homogenous state of strain, as is necessary for the determination of a true material property, material testing is typically done on a system containing the sample with material properties then extracted from the system response.

Although continuous systems have an infinity of vibratory modes, unless the system is highly nonlinear or the modes very closely spaced, the response near a resonant or natural frequency is quite well represented by the single mode dominant at that frequency. And, although the systems of interest may not be truly linear nor is the damping likely to be viscous, we base the analysis of system damping on the well-known responses of the classical one-degree-of-freedom system consisting of a mass, a linear spring, and a viscous damping element.

Our objective is to relate observable measures of system response to the damping present in the system and then to use, with caution, these same measures in the characterization of systems for which the origin of damping is of a more general nature.

A. The Prototype Damped System

The familiar linear oscillator with linear spring, linear Newtonian dashpot, and mass has a displacement x(t) in response to a time dependent force F(t) that is given by:

$$M\frac{d^2x(t)}{dt^2}+C\frac{dx(t)}{dt}+Kx(t)=F(t) \qquad (5)$$

M, C, and K are presumed to be real constants, independent of frequency or amplitude. The solution may be written in terms of two parameters: a natural frequency, $\omega_n=\sqrt{K/M}$, and a fraction of critical damping (damping ratio), $\xi=C/(2\sqrt{KM})$. The homogeneous solution (free vibration)

$$x(t)=C\exp(-\xi\omega_n t)\cos(\omega_d t-\phi_D) \qquad (6)$$

is a decaying sinusoid, shown in Figure 2a for several values of the fraction of critical damping.

The frequency response function, i.e., the magnitude of the response to a harmonic excitation at frequency Ω, is shown in Figure 2b. The displacement is normalized by the response in the limit as forcing frequency goes to zero (static displacement). Values for three levels of damping ratio, ξ = 1%, 2%, and 5%, are given as functions of a dimensionless frequency ratio, $f=\Omega/\omega_n$.

$$R=\left|\frac{x(t)}{x_{st}}\right|=\frac{1}{\sqrt{(1-f^2)^2+4\xi^2 f^2}} \qquad (7)$$

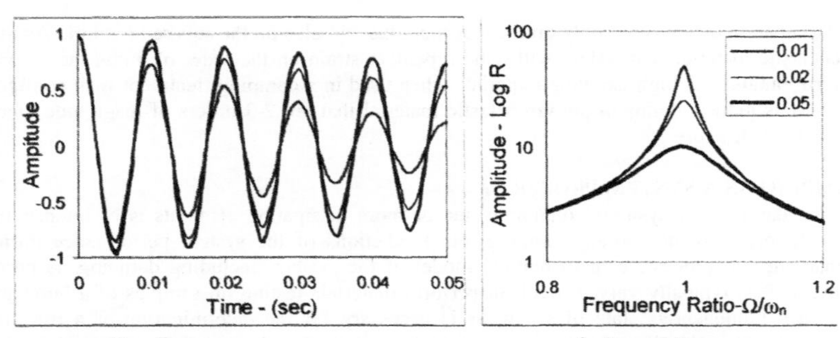

a. Free Response b. Forced Response

Figure 2. Response of Linear 1DOF System at various Levels of Damping

B. Logarithmic Decrement

The logarithmic decrement is defined in terms of the ratio of the amplitude at an arbitrary cycle, n, to the amplitude after an arbitrary number, N, of additional cycles.

$$\delta = \frac{1}{N}\ln\frac{x_n}{x_{n+N}} = \frac{2\pi\xi}{\sqrt{1-\xi^2}} \tag{8}$$

Because the displacement peaks of Figure 2a would fall on a straight line if plotted semi-logarithmically, observations at any pair of cycles, n and n + N, lead to the same result for a linear system. But the presence of a curvature in the time history of the logarithm of amplitude is indicative of a dependence of the decrement on amplitude, i.e., nonlinear damping.

C. Resonant Amplification (or Quality) Factor

The maximum amplitude may be found from the frequency response function, Eq. 7.

$$R_{\max} = \frac{1}{2\xi\sqrt{1-\xi^2}} \equiv Q \tag{9}$$

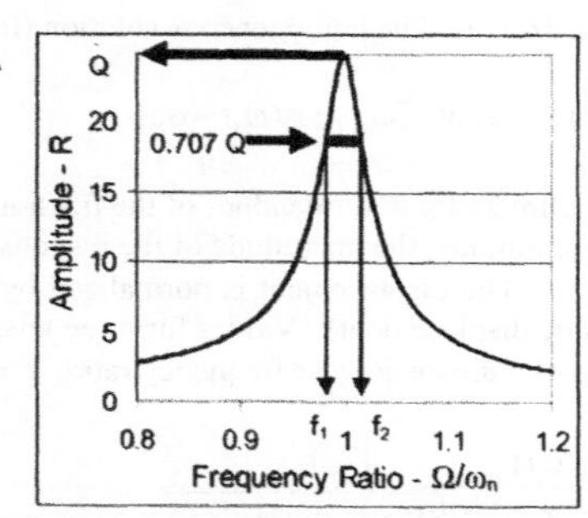

Figure 3. Response near Resonance for a One-DOF Viscous System

The response function shown in Figure 3 is for $\xi = 0.02$. As the maximum depends only on the damping, it provides an observable measure of Q for the system. However, unless the ordinate can be properly scaled by the static displacement, this measure is useful only for comparing the influence of small changes made to the damping of the same system. For small values of damping, $Q \approx 1/2\xi$.

D. Bandwidth for Half-Power Frequencies

Another attribute of the resonant response shown in Figure 3 has come to be the "industry standard" for the determination of the damping of a system. Not only is the peak response determined by the damping, so is the relative width of the response function. Since the energy of a linear system is proportional to the square of the amplitude, the system is said to be at "half-power" at those frequencies for which the amplitude is $1/\sqrt{2}$ of the peak value. There will be two such frequencies, one above and one below the frequency of maximum amplitude, as may be seen in Figure 3. These frequencies may be determined by setting the expression for the magnitude of the response, Eq. 7, equal to $Q/\sqrt{2}$ and solving for the two frequencies, f_1 and f_2, at which that amplitude occurs. For small values of viscous damping, i.e., $\xi \ll 1$, we have $B \approx 2\xi$, since

$$B = f_2 - f_1 = \frac{2\xi\sqrt{1-\xi^2}}{[(f_1+f_2)/2]} \cong 2\xi(1+\frac{5}{2}\xi^2+.....) \qquad (10)$$

E. Evaluation of Dissipation from the Hysteresis Loop

The hysteresis loop, the trajectory of force vs. displacement, is an ellipse for a linear system with viscous damping, and the enclosed area is the energy dissipated, D. In any dissipative system, the maintenance of a sinusoidal displacement-time history requires application of a force-time history out of phase with the displacement. While the hysteresis loop will not be an ellipse in a non-linear and/or inviscid system, the area remains a valid measure of the dissipation. The area under the right triangle connecting the origin to the point of maximum displacement is the peak stored energy in the system, U_S, enabling the specific damping capacity for any system to be evaluated from the ratio of these two measures of energy, $\Psi = D/U_S$. Also, the ratio of the minor to major axes of the hysteresis loop provides an approximate indication of the fraction of critical damping for a general system.

F. Relationships between Measures of Damping

The fraction of critical damping (or damping ratio) has been used as the fundamental measure of system damping. While it is not an observable quantity, we have seen that it may be related to a several observable measures: the logarithmic decrement, the quality factor, the bandwidth, and the specific damping capacity of the system. The first three are implicitly defined at (essentially) the same specific frequency, i.e., at a natural or resonant frequency of the system. Thus, for the same system, they may be readily linked. The system specific damping capacity differs in this regard, however, in that it is applicable at any frequency.

A system loss factor may be defined from the ratio of energy dissipated per cycle to the peak stored energy and then evaluated from any of the observable measures. Because of the likelihood of amplitude dependence, the system loss factors obtained by this means must be regarded as (potentially) dependent on both amplitude and frequency. The relationships are particularly simple for light damping, i.e. $\xi^2 \ll 1$.

$$\eta_S(\omega, X) \equiv \frac{\Delta W(\omega, X)}{2\pi U_S} = \frac{\Psi_S(\omega, X)}{2\pi} = \frac{\delta(\omega_n, X)}{\pi} = \frac{1}{Q(\omega_n, X)} = \frac{1}{B(\omega_n, X)} \qquad (11)$$

And finally, the system loss factor must not be confused with a material loss factor.

G. Adaptation to Other Damping Mechanisms

These same measures of damping may be used with caution as representations of the damping of mildly non-linear systems. Three issues are of particular concern. First, in the presence of non-linear stiffness the period of oscillation of free vibration will vary throughout the decay. In the case of a forced vibration, the frequency response function, Figures 2b and 3, will not be symmetric about the resonant frequency. For sufficiently strong stiffness nonlinearities, the resulting instability may preclude the determination of one of the bandwidth frequencies. Second, in the presence of non-linear (amplitude dependent) damping, the logarithmic decrement will change as the decay progresses and the amplitude diminishes. In consequence, it must be evaluated over a narrow and moving window of cyclic amplitudes. In the case of a forced response, the damping is different at all amplitudes of the frequency response function. In consequence, the bandwidth will not provide a true measure of the system loss factor and must be appropriately adjusted.[4] Finally, a measure of damping given for a nonlinear system is essentially meaningless unless the corresponding amplitude is also provided.

IV. APPLICATIONS FOR ENHANCED DAMPING

Although other means of increasing the passive damping of vibrating systems have been developed (tuned mass dampers, impact dampers, and frictional devices), interest here is in those damping treatments which seek to exploit the inherent damping of a thin layer of a high damping material added to the surface of a component. These fall into two classes, free and constrained layer treatments. The essential distinction is whether the added material is subjected to (essentially) the same strain as the base structure, or if some mechanism for strain amplification is provided.

A. Free Layer Damping Treatments[5]

Free layer damping treatments are constructed by coating some or all of one or both sides of a member in a vibrating structure with a material capable of dissipating energy when subjected to a cyclic strain. A segment of a beam, coated on one side only, is shown in Figure 4. For a coating on one side only, the neutral plane is shifted a distance δ from the neutral plane of the substrate.

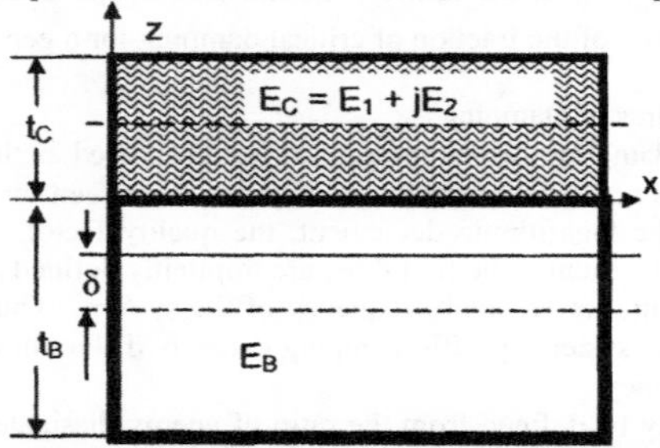

Figure 4. Section of Free-Layer Damping Treatment

Although the coating need not be of a uniform thickness, it will here be treated as such. The development will be outlined for the case of a beam, but the same result is obtained in the case of a plate if the Poisson's ratio for the substrate and coating are the same. It is assumed that the dissipative properties of the coating may be represented by a complex modulus, with the provision that if the modulus is frequency dependent the parameters of the modulus are to be evaluated at the frequency of vibration for the coated structure, which will differ from that of the uncoated structure.

After incorporating the beam and coating moduli, E_B and E_C, and thicknesses, t_B and t_C, into a stiffness ratio and a thickness ratio, $S \equiv E_C t_C / E_B t_B$ and $t \equiv t_C / t_B$, an effective stiffness for the section shown in Figure 4 may be found. For a complex value of coating modulus, the system stiffness becomes complex; a loss factor may be evaluated[3] from it as:

$$\eta_{SYS} = \frac{\Im[(1+St^2)+3S(1+t)^2(1+S)^{-1}]}{\Re[(1+St^2)+3S(1+t)^2(1+S)^{-1}]} \tag{12}$$

While this general result is not conveniently written in as a simple expression in terms of the loss factor of the coating, the result is easily evaluated numerically. Some numerical results are given in Figure 5 for various values of the ratio E_1 / E_B and the thickness ratio t_C/t_B.

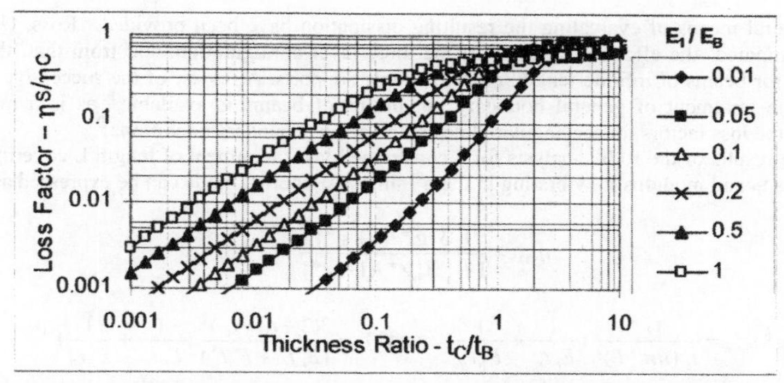

Figure 5. Loss Factors (Normalized) for Free Layer Treatments - Various E_1/E_b

These computations were performed for a coating loss factor of $\eta_C = 1.0$ with the results then normalized by division by that loss factor. The normalized ratio for any other value of material loss factor, η_C, is indistinguishable up to a thickness ratio of 10% and a modulus ratio of unity. The results of Figure 5 are applied by multiplying the value read from the ordinate by the material loss factor of interest.

For small ratios of moduli (as at $E_1 /E_B = 0.01$ of Figure 5) and thickness ratios on the order of unity, as might be characteristic of automotive undercoats, the loss factor is found to vary with the square of the thickness ratio. But for the thin, stiff coatings, as in the left half of Figure 5, a simple result is found:

$$\eta_{SYS} \cong 3\eta_c \frac{E_1 t_c}{E_b t_b} = 3\frac{E_2 t_c}{E_b t_b} \tag{13}$$

B. Constrained Layer Damping Treatments[6]

A constrained layer damping treatment (Figure 6) is formed by bonding a thin elastic sheet to a thin layer of a dissipative material adhering to a structural member. Under a bending deformation, an elastic constraining layer with free ends elongates less than does the surface of the structure and induces a shear strain in the damping layer. By making this layer very thin, the strain is amplified. As many viscoelastic materials can withstand a shear strain of unity, the potential dissipation is very high.

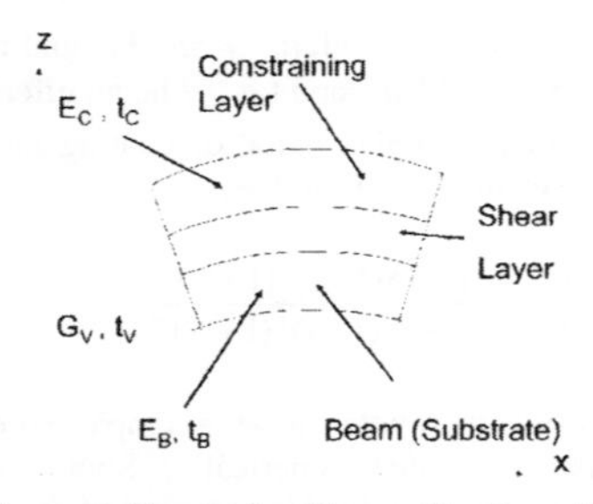

Figure 6. Constrained Layer Configuration

Several means of evaluating the resulting dissipation have been provided. Ross, Ungar, and Kerwin computed[7] the effective stiffness of the three layer configuration, and from that, the system loss factor for beams of infinite length, or with simple supports. A review of the successive advances enabling the treatment of general boundary conditions of beams is available,[8] as is a method for estimating the loss factors for rectangular plates with various boundary conditions.[9]

The results of the RUK analysis for a constrained layer treatment of length L covering a beam of thickness t_B and modulus E_B vibrating in the nth simply supported mode can be expressed as

$$\eta_{SYS} = \frac{Y\,\Im\{g^*/(1+g^*\}}{1+Y\Re\{g^*/(1+g^*\}} \tag{14}$$

$$g^*|_{Beam} = \frac{G^*}{t_V(n\pi/L)^2}\left(\frac{1}{E_C t_C}+\frac{1}{E_B t_B}\right) \text{ and } Y = \frac{3(2+t_S+t_C)^2}{(E_C t_C^3+E_B t_B^3)}\left(\frac{1}{E_C t_C}+\frac{1}{E_B t_B}\right)^{-1} \tag{15a, b}$$

where G* is the complex shear modulus of the damping layer of thickness t_V and the constraining layer is of thickness t_C and modulus E_C. Stretching of the constraining layer reduces the shear strain in the damping layer, thereby reducing the dissipation. In order to maximize the system loss factor, it is necessary that the parameters of the shear and constraining layers satisfy the relationship:

$$\left|\frac{|G^*|L^2}{E_c t_c t_V}\right|_{opt} \cong (n\pi)^2 \tag{16}$$

From this it is seen that optimum damping can not be simultaneously achieved for several modes.

In the usual application, the constraining layer is constructed of a thin metal foil. However, the successful bonding of such a foil to a layer of damping material on a surface with compound curvature, such as an airfoil, is problematic. To address this problem, Rongong, *et. al.*, successfully constructed a constraining layer by plasma-spraying a ceramic layer onto the damping layer.[10]

V. DAMPING OF CERAMIC COATINGS

A. Techniques for Determining Coating Properties

A distinctive feature of sprayed ceramic coatings is that the damping depends on history and amplitude of vibration. System loss factors (η = 1/Q) are shown in Figure 7a for a 1.59 mm super alloy cantilever beam coated on both sides over about 20% of the length in a high strain region with 0.12 mm of 8% yttria stabilized zirconia. System damping was determined[11] from the decay of free vibrations in the second mode from initial amplitudes of 75, 150, and 300 ppm, successively, and adjusted by subtracting the bare-beam loss factor of 0.00033. It is noteworthy that the loss factors observed for all decays do not fall on the same line. A decay initiated at higher amplitude led to higher damping throughout the decay than did a decay initiated at lower amplitude.

A dependence on history was also noted[12] in the determination of system damping in the second cantilever mode of a 152mm long, 2.3 mm thick titanium beam fully covered on both sides with plasma sprayed alumina (0.25mm) on a NiCrAlY bond coat. Values of system loss factors determined by bandwidth at successively increasing amplitudes led to the increasing values of loss factor shown in Figure 7b as solid points. But when retested with values of maximum amplitude monotonically decreasing from the maximum, the measurements denoted with open circles resulted.

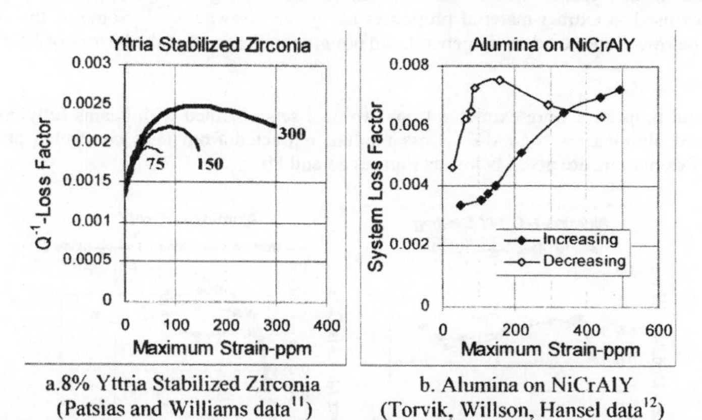

a.8% Yttria Stabilized Zirconia (Patsias and Williams data[11])

b. Alumina on NiCrAlY (Torvik, Willson, Hansel data[12])

Figure 7. Influence of History on System Damping of Beams with Plasma Sprayed Coatings

In order to extract material properties of a non-linear coating material from data for fully covered beams, such as in Figure 7b, account must be taken of the variation in strain throughout the coating. The observed system damping is the integrated contribution of coating dissipation at all strains from zero to the maximum value. For a thin coating, the variation in strain throughout the coating volume is nearly the same as the variation at the beam-coating interface. The uniformity of the strain distribution may be characterized by determining from the mode shapes the fraction of the coating at or below a given fraction of maximum strain, i.e., $A(\varepsilon < \varepsilon_{max})/A_0 = g(\varepsilon/\varepsilon_{max})$. This function varies with boundary condition and mode number but for the first several modes of cantilever, free-free, and simply supported beams it is roughly proportional to the fraction; $g(\varepsilon/\varepsilon_{max}) \approx \varepsilon/\varepsilon_{max}$.

The extraction of the unit damping function requires the solution of an integral equation with the weighting function being the derivative of the area-strain relationship. Given a series of measurements at various levels of maximum strain of the total dissipation in a thin coating of thickness t_C fully covering a beam area A_o, the total energy dissipated is related to specific damping energy by

$$D_{total}(\varepsilon_{max}) = \int_{volume} D(\varepsilon)dV = t_C A_0 \int D(\varepsilon)\frac{d(A/A_o)}{d(\varepsilon/\varepsilon_{max})}d(\varepsilon/\varepsilon_{max}) \qquad (17)$$

One can assume a functional form for the dependence of D (ε) on local strain, and then use the observed values of total dissipation with Eq. 17 to find the parameters of that assumed form. An alternative methodology is to coat only a small area of the beam in a region of nearly uniform strain and then neglect the resulting modest variations in strain. The specific damping energy of the material is then approximated by averaging the total dissipation over the coating volume, or

$$D(\varepsilon) \cong D_{total}(\varepsilon_{max})/(A_o t_c) \quad (18)$$

B. Survey of Coating Properties

A review was recently prepared of the damping properties of hard coatings as found in the open literature. In some cases, material properties were given in the source; in others, material properties were deduced from system measurements. Data sources, original system level data and the methodologies used to extract material properties are given elsewhere.[12,13] Some of the findings are summarized below. Reported data, in general, did not give indications of the history of loading.

1. Alumina

Material properties were extracted from two data sets obtained with beams fully covered with plasma-sprayed alumina on NiCrAlY. Loss moduli, extracted from tests conducted at decreasing amplitudes of excitation, are given below as Figures 8a and 8b.

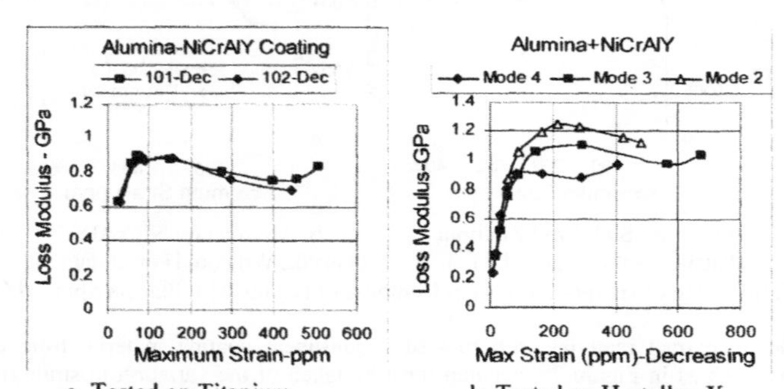

a. Tested on Titanium b. Tested on Hastelloy X

Figure 8. Loss Modulus of Plasma Sprayed Alumina-NiCrAlY Coating System (Data taken by UTC staff using AFRL Facility)

2. Magnesium Aluminate Spinel

Values of the loss modulus for magnesium alumina spinel, extracted from the response at increasing amplitudes of fully covered beams vibrating in the 2nd, 3rd and 4th modes, are shown in Figure 9a. The properties of magnesium spinel were also extracted from vibration decay tests at Sheffield University on partially coated beams of a C263 super-alloy vibrating in the 2nd cantilever mode. Material properties (loss factor and storage modulus) were treated as unknowns at each of 20 values of strain. Trial values of each were then adjusted through iteration so as to produce satisfactory agreement between observed and predicted values of system loss factors and natural frequencies. The loss modulus in Figure 9b was formed from the product of their (unscaled) results. As the system loss factors for ceramic coated beams depend on load history (see Figure 7), material properties extracted from measurements at decreasing levels of excitation can be expected to display the same trends.

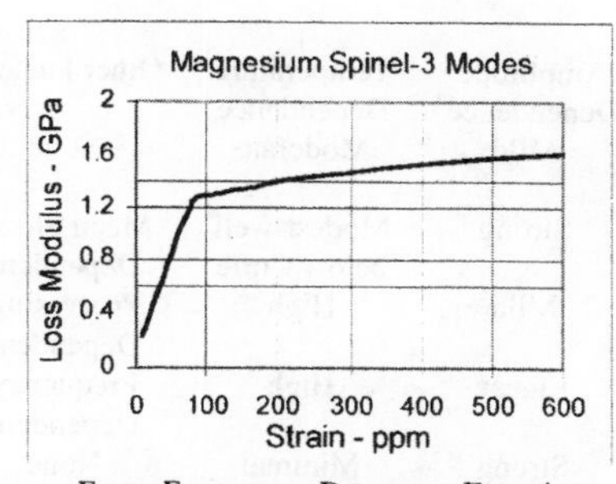

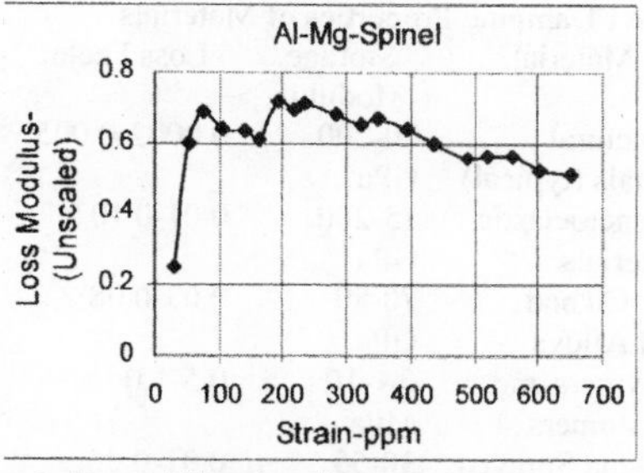

a. From Frequency Response Functions (Data taken using AFRL Facility)

b. From Vibration Decay Records (Using results from Patsias and Williams)

Figure 9. Loss Modulus of Plasma Sprayed Magnesium Aluminate Spinel

3. Rokide® A

Volume-averaged values of the loss factors and storage modulus have also been determined from vibration decay data for fully covered beams. The storage modulus ranged from 30-40 GPa. The strain dependent loss modulus shown in Fig. 10a was constructed from reported results. Beams were coated with Rokide® A aluminum oxide, combustion sprayed at room temperature.

4. Yttria Stabilized Zirconia

Properties of 8 wt% yttria stabilized zirconia (YSZ) deposited by air plasma spray (APS) and electron beam physical vapor deposition (EB-PVD) methods have been compared. In this case, the coating was applied on both surfaces of the beam, but only over 13% of beam length, at the root. No bond coat was used with air plasma sprayed specimens; coatings applied by electron beam physical vapor deposition were on top of a NiCoCrAlY bond coat applied by air plasma spray. Values of the material loss factors and storage modulus were reported; the values of loss modulus shown in Figure 10b were constructed from these.

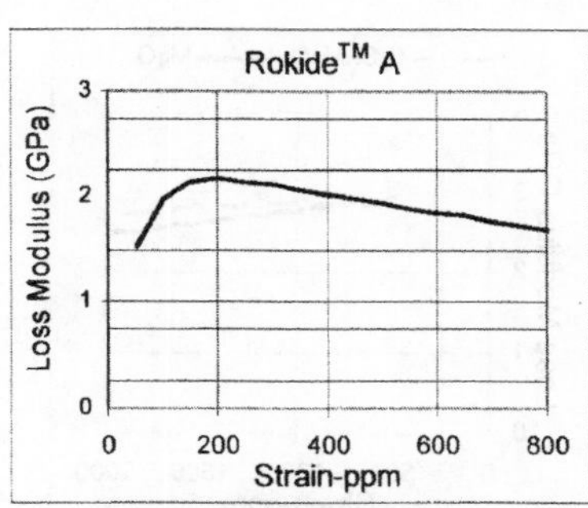

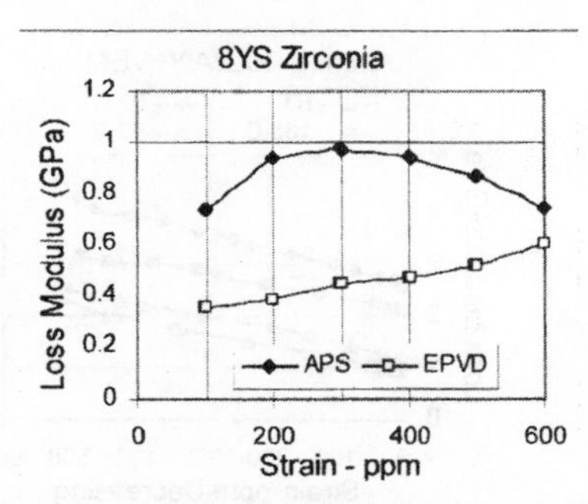

a. Rokide® (Data by Patsias, *et. al.*)

b. 8wt% YSZ (Data by Tassini, *et. al.*)

Figure 10. Loss Modulus of Sprayed Ceramics

Damping properties of sprayed ceramics and other materials are compared in Table I.

Table I Damping Properties of Materials

Material	Storage Modulus	Loss Factor	Amplitude Dependence	Temperature Dependence	Other Factors
Structural Metals (typical)	70-200 GPa	0.0002-0.005	Mild	Moderate	None
Magnetoelastic Materials	15-200 GPa	0.01-0.10	Strong	Modest-well below Curie	Mean Stress Dependent
Mn-Cu and SM Alloys	70-80 GPa	0.03-0.08	Mild	High	Processing Dependent
Polymers and Elastomers	0.1-10 MPa	0.5-1.0	Linear	High	Frequency Dependent
Plasma Sprayed Ceramics	30-55 GPa*	0.01-0.06	Strong	Minimal	None

*Values of the coating storage modulus deduced from the natural frequencies of coated beams.

In the case of a beam or plate covered on both sides, the system loss factor is twice that of Eq. 13. Thus, if the total coating thickness is to be limited to 10% of the thickness of a titanium substrate (i.e., 10% on one side or 5% on each side), the required loss modulus to obtain an increase in system loss factor ($\eta = Q^{-1}$) of 0.01 must be on the order of 3.6 GPa.

This is a daunting requirement, as typical values of the loss modulus for plasma sprayed ceramics appear to be limited to the range of 1-2 GPa. Coating materials with notably higher values of loss modulus (product of storage modulus and loss factor) are necessary.

Impregnation of plasma sprayed ceramic with a viscoelastic component has been considered. It is conjectured that the viscoelastic constituent would then dissipate additional energy through being deformed by the relative motion of the platelets formed in the spray process. Patsias[14] reported that a doubling of system loss factors at 50-150 C resulted from impregnating alumina with polyurethane. In other work,[15] the surface infiltration of a viscoelastic material in alumina led to material loss moduli for the impregnated ceramic shown in Figure 11a and seen to be 300-400% of the values for unimpregnated alumina, Figure 8. While this approach shows promise of leading to the desired values of loss modulus, a frequency dependence not seen in uninfiltrated ceramics can be expected.

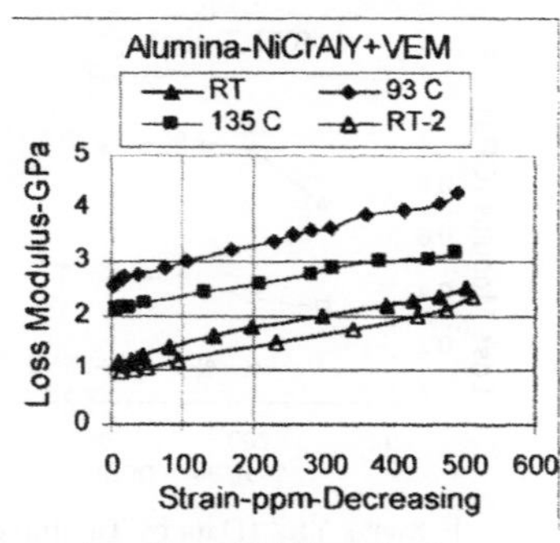

a. Alumina with Viscoelastic Infiltrate (Data by Torvik, *et. al.*)

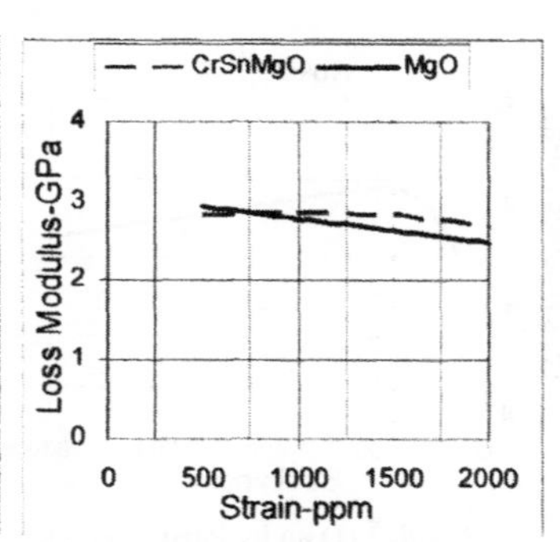

b. EBPVD Graded Coating at RT (Data by Ustinov, *et. al.*)

Figure 11. Enhancements to the Damping of Sprayed Ceramics

The use of electron beam physical vapor deposition to deposit a ceramic on top of a thin layer of a highly deformable metallic layer has also been considered.[16] A 40 μm graded coating was formed of 10 wt % Sn, 15 wt% Cr, and balance MgO. Extraction of material properties[13] from reported system data led to values shown in Figure 11b for the coating system, and for the ceramic alone.

C. ORIGIN OF DAMPING IN PLASMA SPRAYED CERAMICS

Shipton and Patsias[17] examined the defect structure of thermally sprayed MgAl spinel and found a system of horizontally aligned solidified splats with adjacent horizontal splat boundaries that are parallel and well aligned (Figure 12a). The population of vertical defects within splat boundaries subdivided the splats into an array of ordered, parallel-sided plates or 'mobile blocks.' As the vertical cracks were observed to open under strain, Shipton and Patsias inferred that the damping results from the friction on interfaces within and between the 'splats.' Experiments simulating a coated beam by a vibrating beam with segmented and overlapping cover plates showed a dependence of damping on amplitude similar to that seen in coatings. A computer simulation[18] employing springs and Coulomb sliders was also found to predict an amplitude dependent loss factor and storage modulus having the same characteristics as those seen in material properties extracted from experimental data. A system of generally horizontal and vertical defects was also found[17] in plasma sprayed yttria stabilized zirconia (YSZ), as is seen in Figure 12b. But the structure here (and also in plasma sprayed alumina) is notably less well ordered than in the case of MgSpinel coatings. And, as the damping of MgSpinel (Figure 9a) is greater than that of these other materials (Figures 8 and 10b) it is inferred that the higher degree of regularity enables a greater effectiveness of the frictional mechanism, leading to higher damping.

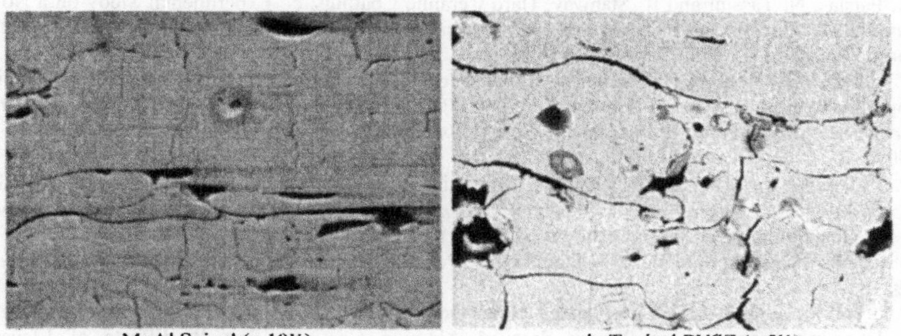

a. MgAl Spinel (x 10K) b. Typical PYSZ (x 5K)

Figure 12. Micro-structure of Typical Plasma Sprayed Ceramics (used with permission)

VI. SUMMARY

The role of damping in the control of vibrations was discussed. Definitions, damping measures, and the damping properties of several classes of materials were summarized and compared, with emphasis given to application as free-layer coatings for the reduction of vibratory amplitudes. The damping potentials of several sprayed ceramics were reviewed and found to be comparable to those of the metallic alloys known to have the highest damping capabilities. The justification for the hypothesis that friction at interfaces between and within the 'splats' resulting from the spray process was also reviewed. The damping capability of plasma sprayed ceramics appears to be less than desired for some applications, although techniques for increasing the damping are being developed. A significant challenge in the characterization of such materials remains, however, as experiments show that the loading history has a significant impact on the damping properties of sprayed ceramic coatings.

VII. REFERENCES

[1]B. J. Lazan, *Damping of Materials and Members in Structural Mechanics*, Pergamon Press, Oxford (1968).
[2]R. L. Bagley and P. J. Torvik, A Generalized Derivative Model for an Elastomer Damper, *The Shock and Vibration Bulletin*, **49**, part 2, 135-143 (1979).
[3]A. D. Nashif, D. Jones, and J. P. Henderson, *Vibration Damping*, John Wiley, New York (1985).
[4]P. J. Torvik, A Note on the Estimation of Non-linear System Damping, *Journal of Applied Mechanics*, ASME, **70**, 449-450 (2003).
[5]P. J. Torvik, Analysis of Free Layer Damping Coatings, *Layered, Functional Gradient Ceramics, and Thermal Barrier Coatings*, ed. M. Anglada, E. Jiménez-Piqué and P. Hvizdoš, *Key Engineering Materials*, **333**, 195-214 (2007).
[6]D. I. G. Jones, *Handbook of Viscoelastic Vibration Damping*, John Wiley, New York, 2001.
[7]D. Ross, E. E. Ungar, and E. M. Kerwin, Jr., Damping of Plate Flexural Vibrations by Means of Viscoelastic Laminae, *Structural Damping*, ed. by Jerome E. Ruzicka, ASME, NY, 49-88 (1959).
[8]P. J. Torvik, The Analysis and Design of Constrained Layer Damping Treatments, *Damping Applications for Vibration Control*, ed. P. J. Torvik, AMD-Vol. 38, ASME, New York, 85-112 (1980).
[9]P. J. Torvik and B. D. Runyon, Estimating the Loss Factors of Plates with Constrained Layer Damping Treatments, *AIAA Journal*, 45, No. 7, 1492-1500 (2007).
[10]J. A. Rongong, A. A. Goruppa, V. R. Buravalla, G. R. Tomlinson, and F. R. Jones, Plasma Deposition of Constrained Layer Coatings, *Proceedings of the Institution of Mechanical Engineers, Part C: Journal of Mechanical Engineering Science*, 669-680 (2004).
[11]A. Patsias, N. Tassini and R. Stanway, Hard Ceramic Coatings: an Experimental Study on a Novel Damping Treatment, *Smart Structures and Materials 2004: Damping and Isolation*, ed. by K-W. Wang, Proc. of SPIE, 5386, 174-184 (2004).
[12]P. J. Torvik, R. Willson, and J. Hansel, Influence of a Viscoelastic Surface Infiltrate on the Damping Properties of Plasma Sprayed Alumina Coatings, Part I: Room Temperature, *Proceedings, Materials Science and Technology 2007 Conference and Exhibition, (MS&T 2007)*, Detroit MI, Sept 15-20 (2007).
[13]P. J. Torvik, A Survey of the Damping Properties of Hard Coatings for Turbine Engine Blades, *Integration of Machinery Failure Prevention Technologies into System Health Management*, Society for Machine Failure Prevention Technology (MFPT), Dayton, OH, 485-506 (2007).
[14]S. Patsias, S., Composite Ceramic Coatings by Polymer Impregnation, *Key Engineering Materials*, **319,** 181-187 (2006).
[15]P. J. Torvik, R. Willson, J. Hansel and J. Henderson, Influence of a Viscoelastic Surface Infiltrate on the Damping Properties of Plasma Sprayed Alumina Coatings, Part II: Effects of Elevated Temperatures and Static Strain, *Proceedings, Materials Science and Technology 2007 Conference and Exhibition, (MS&T 2007)*, Detroit MI, Sept 15-20 (2007).
[16]B. Movchan, and A. Ustinov, Highly Damped Hard Coatings for Protection of Titanium Blades, *Proceedings, RTO AVT Symposium on Evaluation, Control, and Prevention of High Cycle Fatigue in Gas Turbine Engines for Land, Sea, and Air Vehicles*, Seville, Spain, 3-5 October (2005).
[17]M. Shipton and S. Patsias, Hard Damping Coatings: Internal Friction as the Damping Mechanism, *Proceedings, 8th National Turbine Engine High Cycle Fatigue Conference*, Monterey, CA, April 14-16 (2003).
[18]S. Patsias and R. J. Williams, Hard Damping Coatings: Material Properties and F.E. Prediction Methods, *Proceedings, 8th National Turbine Engine High Cycle Fatigue Conference*, Monterey, CA, April 14-16 (2003).

CERAMIC DAMPING COATINGS: EVALUATING THEIR EFFECTIVENESS AND PREDICTING ADDED DAMPING

S. Patsias
Rolls-Royce plc, Experimental Vibration, SIN A-33, PO Box 31, Derby DE24 8BJ,
Tel 00 44 1332 246299, Email Sophoclis.Patsias@rolls-royce.com

ABSTRACT
Ceramic coatings are widely used in Thermal Barrier coating systems to protect the substrates from high operating temperatures. Over the last decade there has been significant research into the potential use of such coatings as damping treatments that has shown that they could provide significant amount of added damping to be of practical value in aero-engine components.

This paper covers the methods developed to (a) evaluate their effectiveness, (b) estimate their material loss factor and modulus of elasticity and (c) predict the added damping when applied to an aero-engine component.

Earlier research has demonstrated that the behaviour of ceramic coatings to be amplitude-dependent (non-linear) and as such, traditional characterisation techniques is rendered unsuitable. A new mixed experimental-numerical approach was developed for such materials: it combines vibration testing and iterative Finite Element modelling that allows one to estimate their material loss factor and modulus of elasticity.

For a new damping treatment to be of practical value one has to demonstrate that by adding extra mass on a component, significant benefits in reducing vibration levels are obtained. This was achieved by predicting the effect of adding a damping coating on a component and by validating the predictions by carrying our experiments of the actual component under static and rotating conditions.

A Rolls-Royce plc propriety software was developed that allows one to add on the finite element model of a component a coating patch of certain thickness at a desired location. This software loads the material properties obtained by the mixed experimental-numerical technique developed earlier and predicts the behaviour of the component.

The experimental validation was carried out with custom made components on a Rolls-Royce plc test rig. Good agreement between experiments and predictions was found and thus validating the predictive capability and proving that ceramic coatings can be of practical application for aero-engine components.

1 INTRODUCTION

The need for additional damping in aerospace components due to the low inherent damping of the base material is well known. Methods that address this need include: friction dampers and Constrained Layer Dampers (CLD) where the induced shear within a soft viscoelastic material constrained by a stiff constraining layer is used to dissipate the energy from the vibration. The main limitation of CLD treatments is the limited temperature range in which the optimum performance lies. Ceramic coatings on the other hand do not suffer from such problems due to their nature.

The use of ceramic coatings as damping treatments has been investigated at the Rolls-Royce plc Materials Damping University Technology Centre (MD-UTC) over the last decade. These coatings provide additional damping in aero-engine components such that vibration levels (and hence the stresses induced) are significantly reduced and component life is extended.

There exist a number of different processes that can be used to form hard coatings. The majority of the coatings are deposited either by Thermal Spraying or by Physical Vapour Deposition. Both methods are widely used for Thermal Barrier Coatings (TBCs), where the first layer, which is termed a Bond Coat can be applied using either process. The top layer tends to be a ceramic based coating, which is usually air (or vacuum) plasma sprayed. Air Plasma Spray (APS) uses Argon and Hydrogen as working gases to melt the sprayed powder particles [1]. The main practical differences between the two methods are the high cost, the limited component size (by the chamber size) and the high temperatures involved (1000 °C) for the PVD case. On the other hand the APS process only reaches temperatures up to 200 - 300 °C (on the substrate) and the size of the component is not as critical.

However, the biggest difference between the two is the microstructure they produce. Figure A1(a) shows a typical APS microstructure. As the powder particles are sprayed on the surface (in molten state) when they reach the substrate's surface they "splat" and form this lamellae structure. On the other hand in Figure A1(b) the PVD microstructure can be observed: it is columnar in nature. Clearly the microstructure is an important parameter in understanding the damping mechanism.

Section 2 of this paper describes the early work on evaluating the damping effectiveness of ceramic coatings, as well as the effect of thickness and temperature. The underlying physical mechanism of energy dissipation in ceramic coatings is discussed in section 3. Previous research has demonstrated that the amplitude-dependent behaviour of hard damping coatings is critical in characterising their behaviour [2]. These characteristics have to be considered, when testing and predicting the added damping and frequency of coated components. Clearly, predictions rely on the quality and accuracy of the material properties they are based upon. A procedure was devised to ensure that the above qualities of material properties were satisfied and it is described in the following section 4. Section 5 describes the Rolls-Royce plc propriety Finite Element prediction routine for ceramic damping coatings. In Section 6 the results for a coated plate tested in the fundamental vibration mode under static and rotating conditions (up to 12,000 rpm) are presented, whereas there results are compared against predictions in section 7. The paper discusses the findings and conclusions in section 8 and further work is discussed in 9.

2 EVALUATING DAMPING EFFECTIVENESS

A number of ceramic coatings were evaluated for the damping capacity they might offer. These included various bond coats and various ceramic oxides. The standard test specimens used for measuring the damping effectiveness of various hard coatings are shown in Figure A2.
For this initial investigation the specimens were excited in the fundamental mode of vibration [3]. These specimens were made of mild steel and were generally coated only on one side. The specimens are clamped in one end and laser displacement probe is used to measure its displacement. The experimental setup can be seen in Figure A3. The signal is recorded on a computer controlling a data acquisition system. The beam is simply displaced to predefined amplitude limited by the allen key and then released: the free decay of the coated specimen in the fundamental mode is stored on the PC. A typical signal and the corresponding FFT spectrum from such tests are shown in Figure A4(a) and (b) respectively.

The damping behaviour from hard coatings was the purpose of a previous study, where it was established that the damping is amplitude dependent [2,4]. That is for different excitation levels the damping levels will vary. To accommodate for this, a Matlab routine was developed to estimate the damping from such decays by employing a moving segment approach. The moving segment method simply averages the damping over each block. This assumes that within the length of the segment the system behaves linearly and hence the damping is estimated using the Logarithmic Decrement (LogDec) method [5]. The length of the segment is selected such that it contains at least 3 cycles. The segment is then moved to a new location and for each point in time the damping is estimated. In Figure A4(a) the first 3 segments are identified as S1, S2 and S3.

The measure of damping for coated components (beams or plates) used in this paper is the Quality factor, defined as,

$$Q = \frac{1}{2\zeta}, \tag{1}$$

where, zeta is the damping ratio based on a viscous damping model, as an equivalent viscous damping approach is utilised here. Also, another terminology that will be used for damping of coated components is the loss factor (eta), defined as,

$$\eta = Q^{-1} = 2\zeta. \tag{2}$$

Therefore, the Q Factors as a function of (Amplitude times Frequency, 'af') are plotted by multiplying the amplitude of the oscillation with the frequency of the oscillation.
Figure A5 shows Q Factors against 'af' level for the signal in Figure A4(a): the first three segments (S1, S2 and S3) are highlighted. Also in Figure A5, the threshold 'af' level above which the damping remains nominally constant is identified as the "linear range", whereas the 'af' level below this threshold is identified as the 'non-linear' range.

2.1 Effect of coating thickness on damping

A ceramic coating was used to coat a number of titanium beam specimens. The coating was sprayed on the top surface only. Two coating thicknesses were used and decay signals were obtained from the fundamental mode of vibration at room temperature.

Figure A6 shows the results for the two different thicknesses as well as the uncoated (reference) specimen. The Q factors have been normalised to the uncoated value, whereas the thicknesses to the maximum coating thickness. The Q factors were obtained using the method described in section 2 and correspond to 'af' levels within the coating's "linear" range.

Figure A6 shows clearly that the damping is proportional to the coating thickness: in other words it follows an approximately linear trend. If one doubles the coating thickness the Q factor will be approximately halved (within the "linear" region).

2.2 *Effect of Temperature on Damping from Hard Coatings*

The effect of temperature on the damping effectiveness of a hard coating was investigated in the early stages of the research [6]. Shouldered specimens were manufactured using Electrical Discharge Machining from Nimonic superalloy bar. The damping level of the uncoated specimen was measured prior to spraying the coating.

A number of specimens were coated with different thickness than the remaining ones, so as to investigate the thickness effect at elevated temperatures. The Q factors were obtained using the method described in section 2. The second flexural mode of vibration was used and the Q factors within the "linear" range were obtained. The results are summarised in Figure A7 where the Q factors have been normalised to the reference (uncoated) specimen. Thickness no.2 is approximately double the thickness no. 1.

Observing the behaviour of the damping added by the hard coating in Figure A7 it is clear that the ceramic coating's damping effectiveness remains nominally constant with temperature. Furthermore the previous finding of the thickness effect was validated by these tests under temperatures up to 600 °C (since the Q factor is halved for double the coating thickness).

Even though the test rig is capable of achieving much higher temperature (1000 °C) it was not possible to conduct the measurements due to the increased damping from the base material at temperatures above 650 °C. It was very difficult to distinguish the added damping due to the coating from the base damping (due to the beam). This is due to degradation of Nickel based superalloys mechanical properties at temperatures between 650 °C and 760 °C [7].

3 DAMPING MECHANISM OF HARD COATINGS

The theory that was formulated in the early stages of this research was that the energy is dissipated by internal friction in the coating [8]. This stems from the fact that the plasma sprayed coatings have a 'splat' or 'pancake' like microstructure, which is formed as the molten powder particles strike the colder substrate surface and rapidly solidify. The proposed damping mechanism was investigated by obtaining a number of Scanning Electron Microscope (SEM) images for a specific hard coating. A typical SEM image is shown in Figure A8, where the main features are highlighted: splats boundaries/interfaces and voids. The voids can be described as porosity or as by-products of the preparation procedure prior to the SEM images. The argument is that the energy is dissipated within the splats boundaries under vibration [9].

Perhaps the most important previous research is described in [10], where plasma sprayed coatings on metal substrates were investigated for the damping capacity of various coatings. They also commented on the amplitude-dependent (non-linear) damping behaviour of these coatings. The authors discuss the bonding of the coating particles (or lamellas) as a combination of fusion and mechanical bonding. One of their statements is "*The dissipation of vibratory energy is, perhaps derived from the friction forces generated at the mechanically bonded interfaces between the coating and the damped structure, as well as at the interfaces with adjacent coating fused areas*". However, this is the extent of their argument, without any micro-structural evidence to support this.

A working model approximating the behaviour of hard coatings was devised by the author: this was based on a simple cantilever beam with attached plastic pads that overlap each other. This can be seen in Figure A10(a) and (b), where a number of overlapping layers of plastic pads are secured through bolts. The beam was then displaced at a set amplitude and released to decay freely – the damping was then estimated from this decay. Also a beam with the equivalent added mass of the dampers was used to provide a useful benchmark for this comparison (shown in Figure A10 (c)).

It is clear that this simple model captures the damping behaviour of the coating quite well: from high 'af' (amplitude-frequency) or strain levels to low it slowly reaches its peak value which then decreases dramatically at lower levels. Furthermore, later research has added to the body evidence, as a preliminary approach to modelling the damping behaviour of ceramic coatings using friction elements [11].

4 MIXED NUMERICAL EXPERIMENTAL PROCEDURE FOR EXTRACTING MODULUS OF ELASTICITY AND LOSS FACTOR OF CERAMIC COATINGS

This procedure has been the outcome of research over the last decade. The need for such a routine becomes clear if one considers the following characteristics of coatings behaviour:

- Amplitude-Dependent damping and stiffness
- Long-Term (or permanent) effects on their properties
- Short-Term (or "memory") effects on their properties.

Figure A11 shows results from coated cantilever beam with 0.2 mm of 8% Y.S. Zirconia, where free vibration was obtained by simply displacing its tip and then allowing the motion to decay freely in its fundamental vibration mode (at 22 ºC) [12]. It is clear the both the damping (plotted as Structural Loss Factor Q^{-1}) and frequency of the coated beam is dependent on the amplitude. In this case, the maximum strain the coating experiences at the coating/beam interfaces is plotted.

Figure A12 shows the results from a number of decays obtained from the same specimen at increasing starting amplitude (or strain): (a) Structural Loss Factor, Q^{-1} and (b) frequency of the coated beam. What is striking is the fact that these decays do not overlap: i.e. there has been a change in the coating's properties – this phenomenon is termed permanent or long-term effect. In Figure A12 (a) and (b) the order the tests were carried out is clearly numbered.

The second feature (termed short-term effect) which is not shown in these graphs, would occur if for example one repeats the decay no. 5 immediately after decay no. 7: one would obtain a different curve

than the original one. It has been found that if coated specimens are left to "rest" for a significant amount of time, the coating will recover its properties.
These interesting phenomena can be easily related to the complex and fascinating microstructure of these coatings, where numerous features control its properties 1[1]. The procedure discussed here was developed such that these features can be accounted in the estimated coating's properties.
The test procedure was briefly discussed in [13], where it was aimed at coated components and not for extracting coating properties. The main difference is that this procedure is a mixture of experimental data and numerical (FE) simulations that allows one to get these properties as a function of the underlying strain. These properties can then be used for predicting the added damping and frequencies of coated components (as discussed in section 5).

The procedure is described in the following sub-sections and a schematic diagram can be seen in Figure A13. The method is based on the Oberst beam technique for extracting modulus of elasticity and loss factor of layered materials on cantilever beams subjected to pure flexure [14]. This procedure is a mixed numerical-experimental method, where the required frequency and damping are measured in vibration tests and the properties are obtained from an iterative Finite Elements Analysis routine. A major difference is the fact that the coating is only applied to a localised area. This is due to the strain dependence nature of the material. Figure A13 shows the geometry and dimensions of the specimen used in this study. The coating is applied to the "high" energy area corresponding to the 2nd flexural vibration mode.

The experimental part of this procedure produces the Q factor (due to the coating) and the frequency of the coated beam at defined 'af' levels. The numerical part uses the above information to estimate the coating's modulus of elasticity and loss factor as a function of the strain at the coating interface (corresponding to the set 'af' level).
The numerical and experimental parts of this procedure are described in the following sub-sections.

4.1 Experimental measurements: frequency and Q factor of coated beams

The test procedure followed here was to measure the natural frequency and damping level of the uncoated beam in the second flexural vibration mode. This was found to be 220.31 Hz and it was measured by a random excitation Frequency Response Function and confirmed by transient decays [5].

Also the background damping, that is damping from the uncoated specimen and any other source of energy dissipation: e.g. air damping was also obtained as a function of the 'af' level. The measurement of damping is not a trivial matter and, when carried out for material characterisation, it is essential to ensure good repeatability of the results. Hence, the damping level for a reference specimen (no coating, made of the same material as the coated specimen) is always measured before any measurements are taken from coated specimens: referred to as the rig loss factor, η_{rig}. If variations in the measured damping exceed 5 %, then the test rig is checked for any faulty equipment or loose and improperly attached specimens or connecting wires.

The measured loss factor from a coated beam is the sum of the loss factors from the coated beam and the energy dissipated in the background (the rig and other factors such as air damping). Therefore, the measured loss factor is corrected, by subtracting the rig damping, as shown in equation (3).

$$\eta_{coated\,beam} = \eta_{measured} - \eta_{rig} \, . \tag{3}$$

As already mentioned, due to the amplitude-dependent behaviour of these materials and phenomena as short-term (memory) effects and permanent changes in their properties, the following procedure was devised:

- A random excitation test was carried out at very low strain level to estimate the approximate frequency of the 2nd flexural mode of vibration.
- A stepped sine test was then carried out, where the excitation level was controlled such that the response amplitude was kept constant at each step (the tolerance on the excitation level was 5%). The frequency range and size of the step was set by trial and error in earlier work with a similar specimen. A typical range could be 223.2 to 223.4 Hz with 101 frequency points (steps of 0.004 Hz, with 5 averages per step).
- The magnitude of the Frequency Response Function (FRF) of a typical stepped sine test is shown in Figure A14 (a). The strain value along the coated length varies by approximately ± 15% from the peak value. From this test, the resonant frequency can easily be obtained as 223.17 Hz.
- The next step is to obtain the damping (Q factor) of the coated beam at this strain level. This is achieved from transient decays, where the beam is excited at the estimated frequency for approximately 3 seconds before the excitation is switched off to record the transient decay.
- The damping is obtained using the technique discussed in [2,15]. The damping is estimated using the envelope of the decay. The Q factor is estimated from the slope of the envelope that covers the first three decay cycles (corresponding to the steady state condition). An example is shown in Figure A14(b). For each 'af' level, five consecutive decays were recorded – so that statistical confidence is ensured. Figure A15 shows the repeatability of the test measurements for five consecutive decays: (a) 1.07% variation in Q Factors and (b) 0.03% variation in frequencies.

The above two steps are repeated for each 'af' level required in an increasing amplitude manner. This is necessary to avoid any permanent changes of the coating's microstructure affecting the results (as already discussed).

4.2 *F.E. iterative routine to calculate coating's Young's modulus of elasticity (E) and loss factor (η)*

This step requires accurate FE models of the uncoated and coated test specimens: the thickness of the uncoated and coated beams is crucial as it affects the accuracy of predicted natural frequencies and hence the accuracy of the estimated material properties. These non-linear properties (modulus of Elasticity and material loss factor) are termed as a Master curve, since this terminology is widely used to characterise viscoelastic materials [14].

The master curve for this work was assembled from a C263 superalloy specimen with an Air Plasma Sprayed ceramic coating. The thickness of the bare beam was found to vary along its length: this variation was included in the FE model of the bare beam. Figure A16 shows the different areas and the various thickness values. The FE mesh was developed with 20-noded brick elements.

The second flexural mode of vibration was calculated using a Rolls-Royce plc propriety software as 220.30 Hz, that is a difference from the experimental value of 0.005%, which is acceptable. The coated beam FE model with the coating on each side of the beam was created using 20-noded brick elements [16].
The main function of the iterative FE routine, is to establish the coating's Young's modulus of elasticity (*E*) at each different 'af' level tested. This was implemented with a Matlab script that modifies the assumed *E* value of the coating, saves the new model and calls a FE routine that runs the eigenvalue analysis for the second flexural mode.

The frequency is then extracted from the results files and compared against the experimental value. Then the *E* value is modified accordingly and the revised model is run. This procedure is repeated until the solution converges to two decimal points of precision.
Once the solution converges, the Matlab routine extracts the Modal Strain Energy (MSE) Ratio [14] so that the loss factor of the coating can be estimated, since [14]

$$MSE\ Ratio = \left[\frac{U_{coating}}{U_{coating\ and\ beam}} \right], \tag{4}$$

$$(\eta_{coating}) = \left[\frac{\eta_{coated\ beam}}{MSE\ Ratio} \right]. \tag{5}$$

4.3 Master Curve for an APS ceramic coating

Using the above two steps of the procedure, a Master Curve was produced for an APS ceramic coating. The results are presented in Figure A17 (a) and (b) for the modulus of elasticity and loss factor respectively. The actual values obtained from the above routine are represented by the blue squares, where the red dotted lines are the fitted curves. These curves were used to avoid any convergence problems with the FE prediction routine for damping coatings (section 3) due some "abnormal" points at the low strain range. The values for both the loss factor and modulus of elasticity have been normalised. The modulus of elasticity data points, were fitted with a logarithmic fit and the loss factor data points were fitted with a rational function [17].

5 FE PREDICTION ROUTINE FOR DAMPING COATINGS

The prediction routine for damping coatings is software plug-in for a Rolls-Royce plc proprietary FE program that predicts damping and frequencies of components with viscoelastic and ceramic coatings as damping treatments.

The FE routine for damping coatings is capable of predicting frequency and damping for components with amplitude-dependent material treatments such as ceramic coatings. The procedure one has to follow when predicting the damping and frequency of coated components is briefly described here. The F.E. model of the component under consideration might already have a coating layer or it can be added in some other way: in any case the coating is modelled by adding elements to the FE model of the component.

The required inputs are: the 'af' level and vibration mode of interest, the tolerance of convergence, the maximum number of iterations allowed and the number of domains into which the coating elements will be divided. A typical analysis setup can be seen in Figure A18(a). The role of each of the above parameters becomes clear, by explaining the iterative routine that FE prediction routine for damping coatings employs to obtain the added damping and frequency at the required 'af' level:

- The first run sets the Young's modulus of elasticity of all the coating elements at the average value and obtains the first approximation of the frequency.

- Based on the average strain in the coating elements it sets the parameters for the second iteration: it divides the range of strain values in 20 domains in this case (as shown in Figure A18(b)) and then assigns the elements that fall in each domain the equivalent modulus of elasticity from the master curve. Figure A19 shows this effect by highlighting the different element domains with different colours. In this case, the fundamental mode of vibration is of interest and hence the strain energy is concentrated near the root (since clamped-free conditions exist).

- The second iteration gives estimates of both frequency and damping: Figure A20 shows the results from such a run (note that the first iteration estimates only the frequency).
- The third iteration will re-assign modulus of elasticity and loss factor values to the coating elements based on the outcome of the second iteration.

- The total number of iterations depends on the complexity of the model and the coating's master curve.

The damping is obtained using the Modal Strain Energy (MSE) method [14] the only variation is that the summation of the contributions of all the coating domains is calculated (since each domain would have different modulus of elasticity and loss factor). Finally, the process is repeated until the convergence accuracy set by the user is met.

6 TESTS OF A COATED PLATE AT THE ROTATING DAMPING RIG

The experimental work into ceramic damping coatings (prior to these results) has been restricted to static vibration rigs, and the need for a simple rotating test facility has been identified. This was used to demonstrate the effectiveness of the various damping technologies under conditions of high centrifugal forces acceleration and the associated steady stress field, and would also enable the integrity of the applications to be explored. The developed rig is used for the work described in this paper.

The rig was developed from an existing vacuum spin rig facility, which had previously been used for rotating vibration measurements. This is a self-contained rig, with the drive motor, oil system, drive shaft and vacuum chamber mounted into a compact frame (Figure A21(a)). The rig has a maximum operating speed of 12,000 rpm (44,000 g). A new rotating assembly was developed for the rotating damping measurements. This is in the form of a transverse beam with mounting features for a specimen and an opposing balance mass (Figure A21(b)). Permanent instrumentation wiring has been bonded to the forward surface of the rotor (carrier) with connection points adjacent to the central bore and to the specimen mount. Flying leads connected to the inboard terminals, pass through the centre of the shaft to an externally mounted 12-channel miniature IDM slip ring unit. This is used to pass dynamic data from the specimen instrumentation, and also to transmit electrical power to the specimen-mounted excitation system.

An essential requirement of the specimen mount and the rig in general, is that it should provide good vibration clamping (i.e. low damping) without being overly complex. This has been achieved by machining a 'shouldered' specimen with a thicker, parallel root. The specimen root tang locates in a keyhole-relieved slot in the carrier, the mating surfaces of which are accurately machined to provide a good sliding fit. An array of six through-bolts clamps the specimen into the slot, providing good vibration clamping. However, neither the bolts nor the resulting interface friction are intended to support the centrifugal load of the specimen. A single 20 mm diameter dowel pin through the centre of the tang provides this function. Whilst there is some clearance enabling the dowel to move slightly, this is not observed to increase damping in the specimen modes, and the dowel 'locks up' under centrifugal load at low speed. An additional factor to be considered is the non-linearity of ceramic coatings. Previous results indicated that damping properties are highly non-linear at low amplitudes, but tend to plateau at amplitudes above 0.1 mHz [15]. A more important base line criteria, therefore, is that it must be possible to excite the specimen beyond any threshold value for the material properties. The excitation system deployed was based on Piezoelectric actuators attached on the specimens (as shown in Figure A21 (b)).

Due to the amplitude-dependent behaviour and the memory effects (both long-term and short term) a procedure was devised to ensure that the damping properties were measured systematically and any of these effects were quantified:

- Firstly a series of transient measurements were recorded for a range of increasing initial amplitudes, including very low amplitudes. Care was taken to ensure that the specimen was not exposed to any high amplitude conditions out of sequence, and a short time for the properties to stabilise was allowed for before each measurement.
- The specimen was then held at the high amplitude condition (0.2 mHz), and a further measurement recorded to measure the effects of extended time on condition.
- The specimen was allowed to rest for an hour before the incremental test was repeated. This second test identifies any permanent property changes occurred during the first incremental test. It was shown that the properties would not change significantly during subsequent incremental tests, providing the maximum amplitude was not exceeded, and the second incremental test therefore provides a reliable datum measurement of the coating properties prior to the rotating tests.
- A final set of incremental measurements would then be made following tests under rotating conditions to establish whether any further change in properties had taken place.

It was not feasible to repeat the full test sequence described above under rotating conditions due to the extensive duration of the testing and the wide range of speeds required. Strain-gauge noise was also much higher under rotating conditions, and this limited the resolution of low amplitude measurements. It was therefore decided to conduct the rotating tests only at the highest amplitude tested statically (~

0.2 mHz), and to repeat this measurement at incremental speeds up to the maximum speed of the rig (12000 rpm). Over this speed range, the fundamental frequency increases by ~ 40% due to CF stiffening, and this would result in a corresponding reduction in the test amplitude if the 'af' level was maintained. Given that the coating properties are strain dependent and are not frequency dependent, it is more appropriate to test to constant amplitude conditions. The higher speed tests were therefore conducted at the same indicated strain-gauge output as the 0.2 mHz static test.

All tests were conducted in vacuum (80 mm-Hg), and the specimen was run on condition for 10 to 15 minutes at the low speed condition, or until the response was stable, before each acceleration.
For each test, a decay was obtained by switching the excitation frequency from the resonant to a higher value: this produced a much smoother transition, with a negligible effect on the response. A curve-fitting algorithm was developed for determining non-linear damping and frequency properties from experimental data contaminated with additional frequency components. For cases where the initial part of the decay waveform was contaminated by switching noise, the software provides a facility for extrapolating the curve fit back to the start of the decay.
The following section presents the results obtained from this rig for the fundamental flexural vibration mode and compares them against predictions from the FE prediction routine for damping coatings.

7 COMPARISON OF PREDICTED AGAINST MEASURED RESULTS FOR MODE 1F

The FE mesh was developed using 20-noded brick elements and the PZT patches (used to excite the blade) were also modelled. It has been found (by earlier work) that these contribute significantly to the blade's stiffness and strain distribution at the coating/blade interface. Figure A22(a) shows the FE model on the coating side and (b) shows the reverse side with the PZT patches highlighted in red.

The coating layer was then added using 20-noded brick elements (as the plain plate) [16]. The coating thickness was measured at 12 locations and it was found that it varied along the length and width. This can be seen in Figure A23(a) where the values have been normalised. As previous research has shown that the added damping from hard coatings is proportional to its thickness [15], this was also included in the FE models. Figure A23 (b) shows a measure of the contribution to damping for the fundamental mode of vibration of this specimen. As the added damping depends on the amplitude and the coating thickness, this needs to be taken into consideration if accurate predictions are to be made. Figure A23 (c) shows the modeshape of the fundamental vibration mode.

The number of elements (for plate, coating and PZT) in the FE model was 2685. The plate material was titanium 6-4. All of the predictions were carried out on an Ultra 60 single processor Sun Workstation with 1.7 GBytes of memory.

The damping is represented as the Q factor (system): for the predictions the measured background damping was added to predicted value. This allows for direct comparisons to be made. Three sets of results (all for the fundamental mode) are used to validate the predictions in this report and discussed in the following sub-sections.

7.1 Static tests at various 'af' levels

For these results under static conditions it is clear that the predictions follow very well the measured data. Figure A24(a) compares the damping values where Figure A24 (b) the compares frequencies at the various 'af' levels tested. It is quite clear that the predicted frequencies are correlated very well with the measured values (average variation of 0.14%). The damping values show generally a good correlation with the measurements, less than 10% at the higher 'af' levels.

The trends shown in Figure A24 demonstrate that the prediction routine and the mixed numerical-experimental procedure for obtaining Master-Curve provide reliable predictions for coated components.

7.2 Rotating tests from 300 to 12,000 rpm, at approximately 0.2 mHz

For these sets of results under rotating conditions, the temperature variation during an increase in rotating speed from 300 to 12,000 rpm was taken into consideration.

Figure A25(a) and (b) show a comparison of predicted and measured results for damping and frequencies under rotating conditions, respectively. The close correlation of the frequency and damping parameters indicates that, the rotational effects have been captured in the model. The scatter in the measured results at high speed has been attributed to secondary excitation resulting from the proximity of the test condition to the rotor critical speed. Also, one has to bear in mind that the FE mesh resolution becomes more important at higher rotational speeds.

7.3 Repeat static tests

Once the rotational damping tests were complete, static damping measurements were repeated. Figures A26(a) and A26(b) compare the measured results before and after the 12,000 rpm run. It is quite clear that there has been a positive change in the damping performance of the coating, i.e.:

- The damping capacity of the coating has increased (lower Q factors)
- The stiffness of the coating has decreased (lower frequencies).

It is noted that these changes are not captured at this time in the prediction code, as can be seen in Figure A27(a) and A27(b) for the Q factors and frequencies respectively. This change is an improvement in the coating material properties: the prediction code should therefore under-predict the damping levels and hence it is acceptable as a design tool.

It is important to emphasise that tests were extended to a much lower 'af' levels, than in the first set of tests. For these low 'af' tests the maximum strain the coating undergoes is at the lower end of the Master-Curve and the algorithm that divides the elements into domains (based on the strain value) is not reliable.

8 DISCUSSION AND CONCLUSIONS

This paper discussed the work carried out or led by the author over the last decade into ceramic damping coatings. The key features of such damping materials are:

- **Amplitude – dependent elastic and damping behaviour**
- **Damping effectiveness when applied on a component can be increased by increasing the coatings' thickness**
- **Damping effectiveness seems to be independent of temperature in the range of 20-600 °C.**

The mixed-numerical-experimental procedure for obtaining Master-Curve has been demonstrated for a ceramic coating.

The FE prediction routine for such materials has been validated against measured data from a ceramic coated plate:

- **Static tests prior to rotating tests at varying 'af' levels**
- **A set of rotating tests at 0.2 mHz and rotating speeds of 300 to 12,000 rpm.**

It has been demonstrated that the coating damping properties are improved when subjected to rotational loading

These improvements are not captured in the Master-Curve and the FE prediction routine for damping coatings will always under-predict the damping (which is acceptable for a design tool).

Reasons for variations between predictions and measurements include:

- **For low 'af' levels, the strain levels will be at the lower end of the Master Curve, this will not allow for the division of the coating element according to the strain to be successful**
- **The Master-Curve was obtained from a C-263 superalloy specimen, whereas the predictions were made for a titanium 6-4 plate. The two different materials could affect the coating built-up (especially at the coating/substrate interface)**
- **Damping measurements are always challenging, especially for amplitude-dependent ceramic coatings: significant variations in measured values can be expected.**

9 SUGGESTIONS FOR FURTHERWORK

Further work on the Master-Curve and validation of the FE prediction routine for damping coatings includes:

- **Extending the validation work to higher modes such as the 1st Torsional mode**
- **Investigation of the substrate material effect on the properties of the ceramic coatings**

- Extension of the strain range of the Master-Curve: both for lower but most importantly for higher strain levels.

Finally, improvements of the FE prediction routine for damping coatings include:

- Use of adaptive FE mesh that allows for optimum division of coating elements
- Variable "length" of strain range for dividing the Master-Curve and optimum number of domains
- Effect of the FE mesh resolution on the accuracy of the predicted damping and frequency values.

10 REFERENCES

[1] Pawlowski, L., *The Science and Engineering of Thermal Spray Coatings*, John Wiley & Sons Ltd, England, UK, 1995.

[2] S. Patsias and R. Williams, "Hard damping coatings: material properties and F.E. prediction method", *8th National Turbine Engine High Cycle Fatigue (HCF) Conference*, Monterey, California, CD-ROM, 12 pages, 14th – 16th April, 2003.

[3] R. D. Blevins, *Formulas for Natural Frequency and Mode Shape*, Krieger Publishing Company, Malanar, Florida, 32950, USA, 1979.

[4] S. Patsias, G. R. Tomlinson and A. M. Jones, "Initial studies into hard coatings for fan blade damping", invited paper, *6th National Turbine Engine High Cycle Fatigue (HCF) Conference*, Jacksonville, Florida, CD-ROM, 11 pages, 5th – 8th March, 2001.

[5] W. T. Thompson, *Theory of vibration with applications*, George Allen & Unwin, 1981.

[6] S. Patsias, G. R. Tomlinson and A. M. Jones, Initial studies into hard coatings for fan blade damping, invited paper, *6th National Turbine Engine High Cycle Fatigue (HCF) Conference*, Jacksonville, Florida, CD-ROM, 11 pages, 5th – 8th March, 2001.

[7] Metals Handbook, Vol.17, *Nondestructive evaluation and quality control*, 9th edition, ASM International, 1989.

[8] Zolotukhin, I. V., Ahkinin, K. G., Abramov, V. V., Netusov, Iu. K., Skorobogatov, V. S. and Svedomtsev, N. V., Investigation of the damping and elastic characteristics of plasma-sprayed tungsten, nichrome, zirconium dioxide, and chrome-Nickel spinel coatings, Journal: Problemy Prochnosti, Vol. 5, pp. 86-89, September, 1973.

[9] M. Shipton and S. Patsias, "Hard damping coatings: internal friction as the damping mechanism", *8th National Turbine Engine High Cycle Fatigue (HCF) Conference*, Monterey, California, CD-ROM, 10 pages, 14th – 16th April, 2003.

[10] K. R. Cross, W. R. Lull, R. L. Newman and J. R. Cavanagh, "Potential of graded coatings on vibration damping", *Journal of Aircraft*, vol. 10, no.10, Nov., 1973, pp. 689-691.

[11] J. Green and S. Patsias, "A preliminary approach for the modelling of a hard damping coating using friction elements", invited paper, 7th National Turbine Engine High Cycle Fatigue (HCF) Conference, Palm Beach Gardens, Florida, CD-ROM, 9 pages, 14th – 17th May, 2002.

[12] S. Patsias, N. Tassini and R. Stanway, "Hard ceramic coatings: an experimental study of a novel damping treatment", Proc. SPIE 5386, Smart Structures and Materials 2004, Damping and Isolation, San Diego, California, K.-W. Wang Ed., pp. 174-184, 14th – 18th March, 2004.

[13] P. J. Torvik, S. Patsias and G. R. Tomlinson, "Characterising the damping behaviour of hard coatings: comparisons from two methodologies", 7th National Turbine Engine High Cycle Fatigue (HCF) Conference, Palm Beach Gardens, Florida, CD-ROM, 17 pages, 14th – 17th May, 2002.

[14] Garibaldi, L and Onah, H.N., *Viscoelastic material damping technology*, Becchis Osirides s.r.l, Torino, June, 1996.

[15] S. Patsias, G. R. Tomlinson and A. M. Jones, "Initial studies into hard coatings for fan blade damping", *6th National Turbine Engine High Cycle Fatigue (HCF) Conference*, Jacksonville, Florida, CD-ROM, 11 pages, 5th – 8th March, 2001.

[16] Spencer A, "Fella – Finite Element Layer Analysis", User Manual, May, 2002.

[17] Hyams, D. , "CurveExpert", User Manual, v. 1.37, 2001.

11 ACKNOWLEDGEMENTS

The author wishes to thank Rolls-Royce plc for giving permission to publish this work. Thanks also to Rolls-Royce plc colleague R. Williams for carrying out the rotating rig testing and all the author's ex-colleagues at the University of Sheffield, where the majority of this work was carried out under a RR research contract.

12 FIGURES

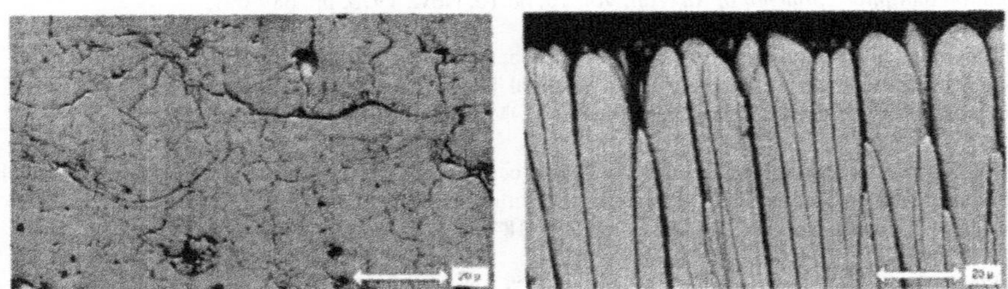

Figure A1: Typical microstructures of ceramic coatings. (a) Air Plasma Sprayed and (b) Electron Beam Physical Vapour Deposition.

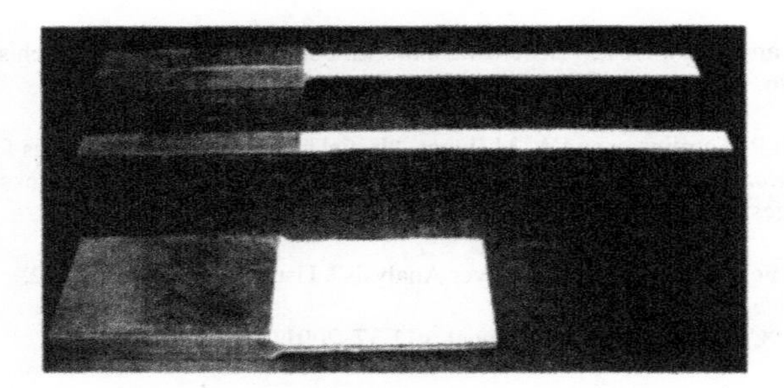

Figure A2: Standard test specimens used to evaluate and characterise ceramic coatings.

Figure A3: The experimental setup for evaluating ceramic coatings in the fundamental mode of vibration.

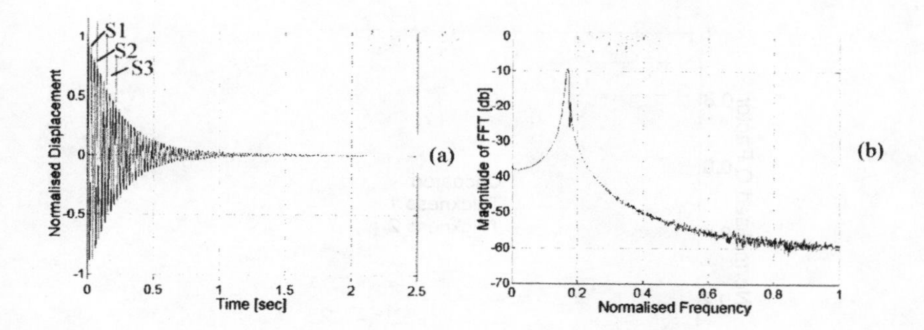

Figure A4: (a) Typical decay of a coated cantilever in its fundamental mode of vibration and (b) the corresponding magnitude of its FFT. Also the first three segments used to extract the damping as highlighted S1, S2 and S3.

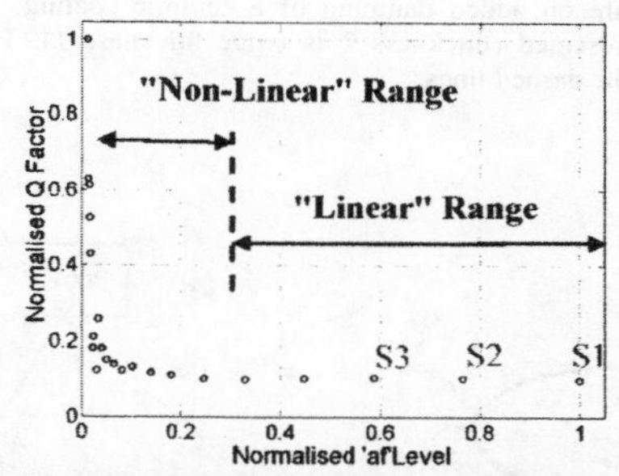

Figure A5: Estimated damping for the decay shown in Figure A4(a). Q factors are plotted against 'af' level for each segment. Segments 1 to 3 are highlighted (corresponding to Figure A4(a)).

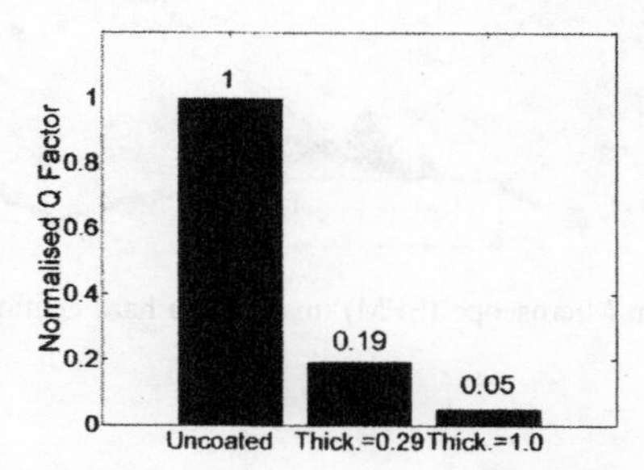

Figure A6: Effect of coating thickness on damping: results from titanium specimens coated on the top face and tested at the fundamental mode of vibration under room temperature.

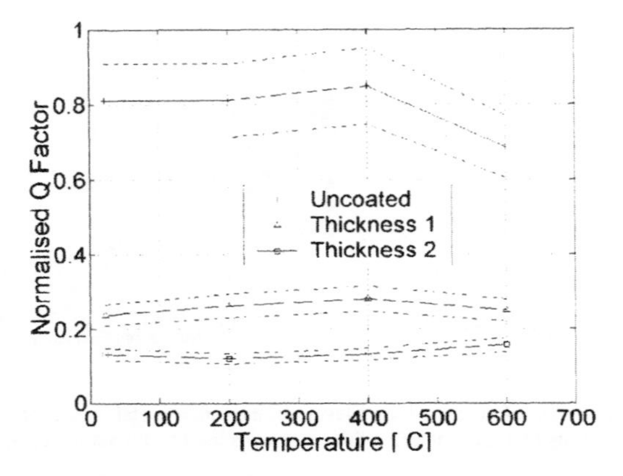

Figure A7: Effect of temperature on added damping of a ceramic coating. The effect of thickness under elevated temperatures is also presented (thickness 2 is twice thickness 1). The experimental discrepancies allowed was 15 %, indicated by the dashed lines.

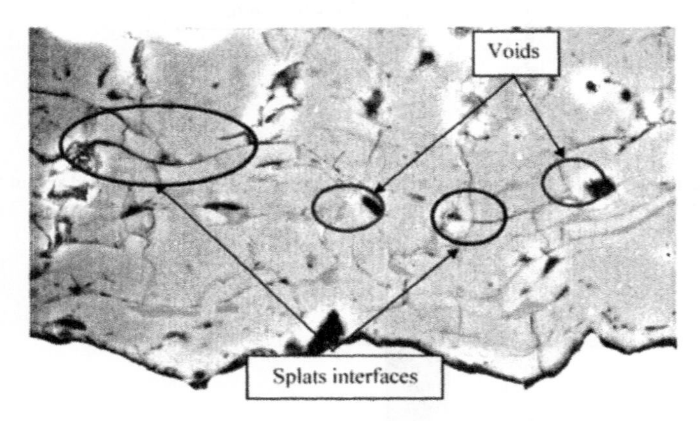

Figure A8: Scanning Electron Microscope (SEM) image for a hard coating – Magnification factor of 2000 was used.

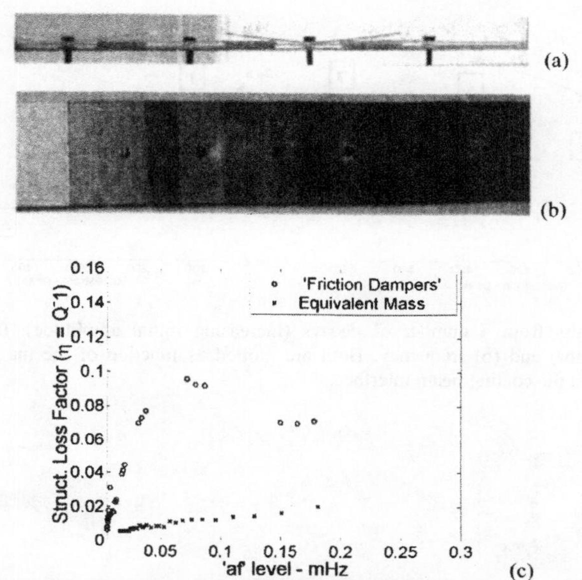

Figure A10: (a) side view and (b) plan view of the "friction dampers" model. (c) Damping behaviour approximates closely that of ceramic coatings

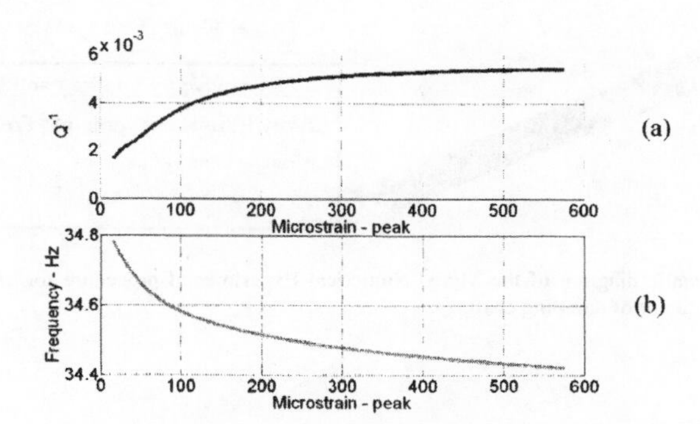

Figure A11: Amplitude dependent behaviour from a beam coated with 8% Yttria Stabilised Zirconia: (a) Damping Q^{-1} (structural loss factor) and (b) frequency. Both are plotted as function of the maximum strain the coating undergoes at the coating/beam interface.

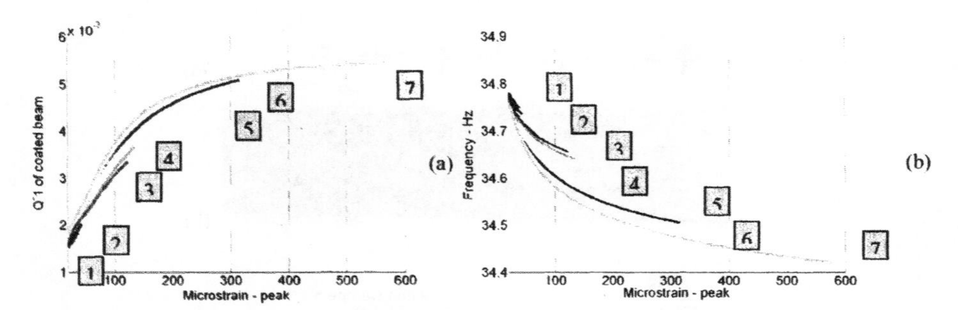

Figure A12: Results from a number of decays (increasing initial amplitude): (a) Damping Q^{-1} (structural loss factor) and (b) frequency. Both are plotted as function of the maximum strain the coating undergoes at the coating/beam interface.

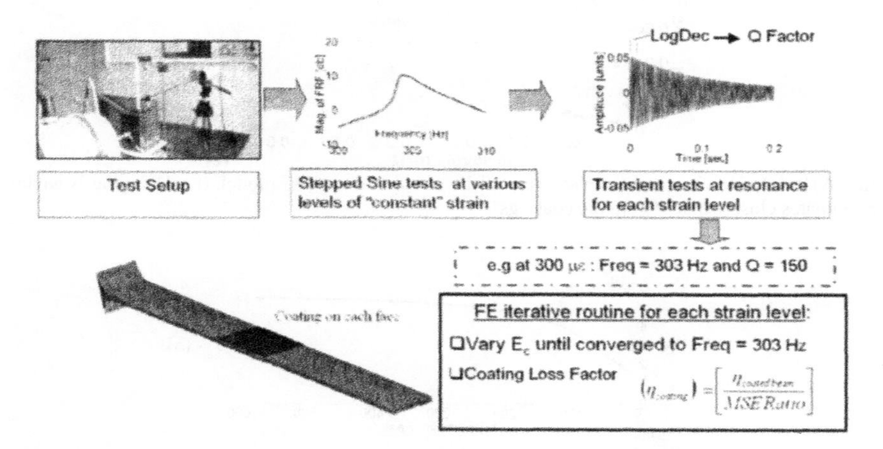

Figure A13: Schematic diagram of the Mixed-Numerical Experimental procedure for extracting Modulus of Elasticity and Loss factor of damping coatings.

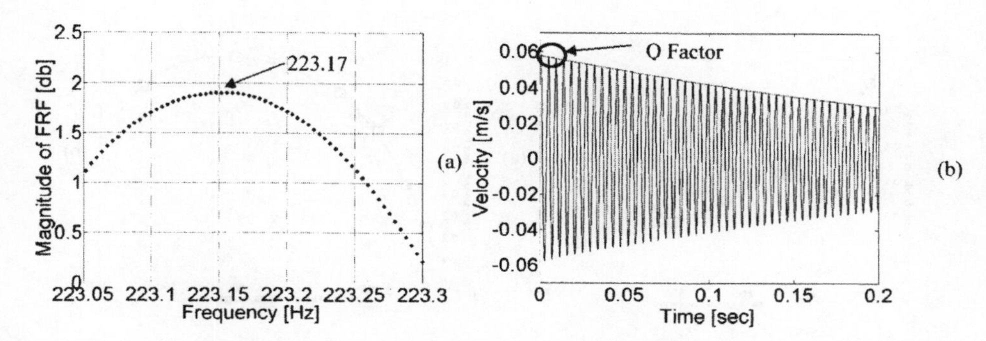

Figure A14: Typical test data from a coated beam used in the procedure shown in Figure 3: (a) typical FRF from a stepped sine test at "constant" strain (b) typical decay at 223.17 Hz and same strain as in (a).

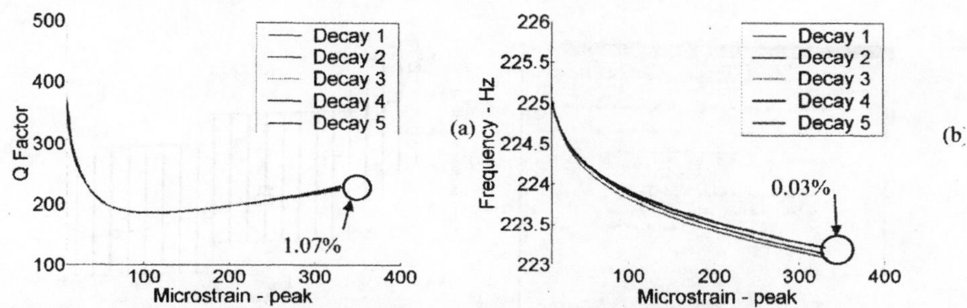

Figure A15: Repeatability of five consecutive decays of the characterisation procedure: (a) 1.07% in Q Factors and (b) 0.03% in frequencies. Both evaluated at the highest strain level (or 'af' level).

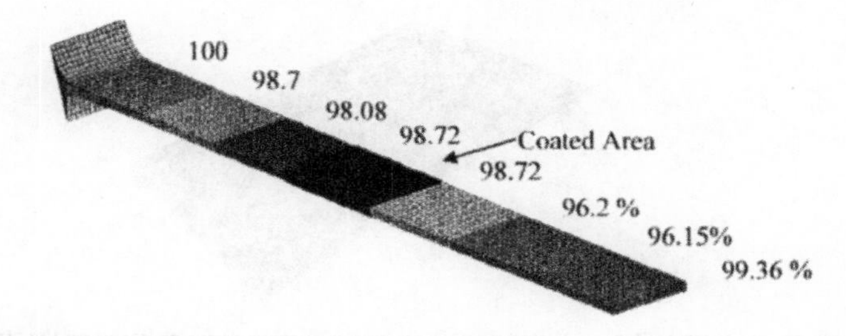

Figure A16: The FE mesh of the uncoated beam: the variation of the thickness of the beam was modelled.

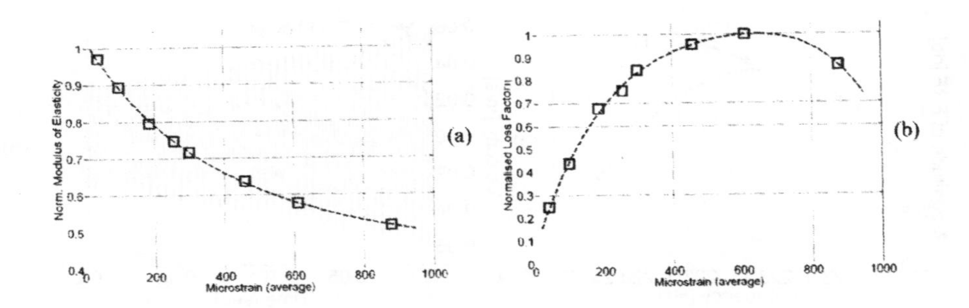

Figure A17: Master-Curve of a ceramic coating: (a) Modulus of Elasticity and (b) Loss Factor.

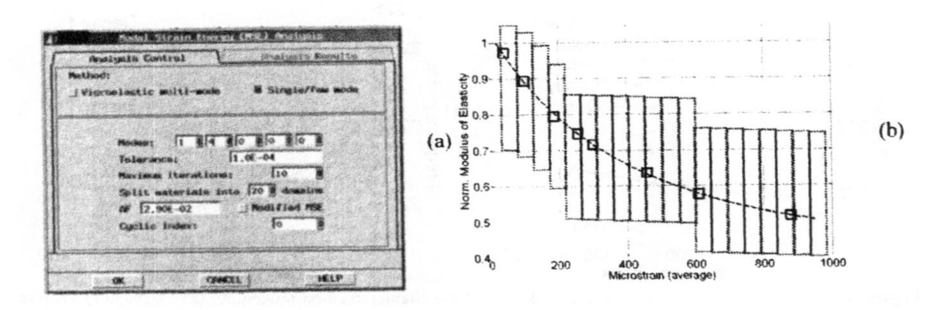

Figure A18: FE prediction routine for damping coatings: (a) analysis setup for hard damping coatings run (b) dividing the ceramic coating's Master Curve into 20 domains.

Figure A19: Example of the FE prediction routine assigning coating domains based on the strain in the coating elements: fundamental vibration mode.

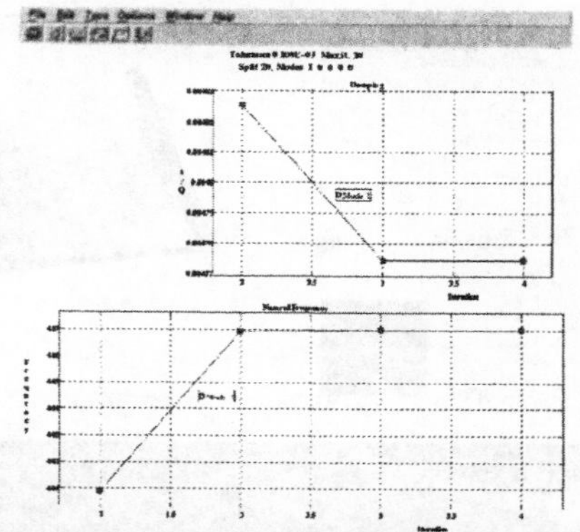

Figure A20: FE prediction routine for damping coatings: iterations convergence of natural frequency and damping after four runs.

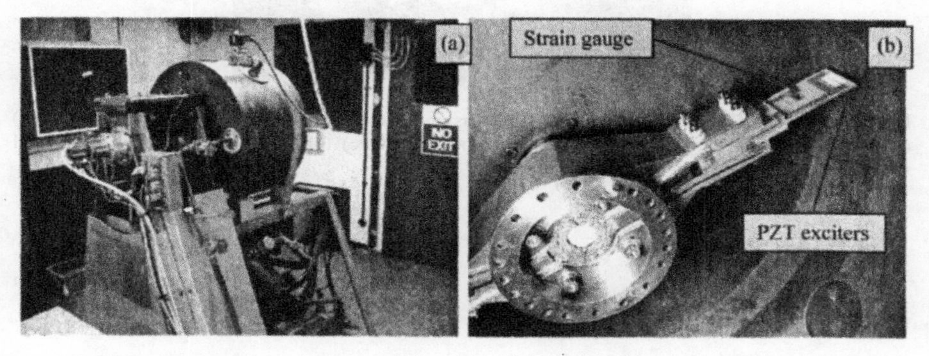

Figure A21: Facility used for testing the coated plates at Rolls-Royce plc. (a) rear view of the vacuum spin rig (b) instrumented specimen carrier assembly. N.B. the specimen shown here was used for other tests (not discussed in this paper).

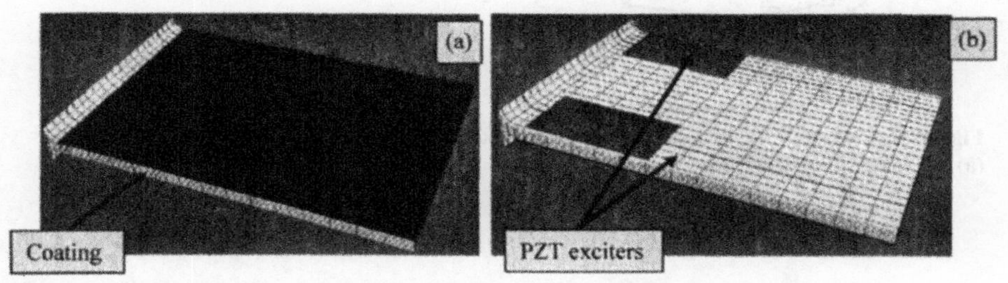

Figure A22: FE model of the coated plate used in this paper: (a) top face where the coating is highlighted in blue (b) the opposite (uncoated) face where the PZT excited are indicated.

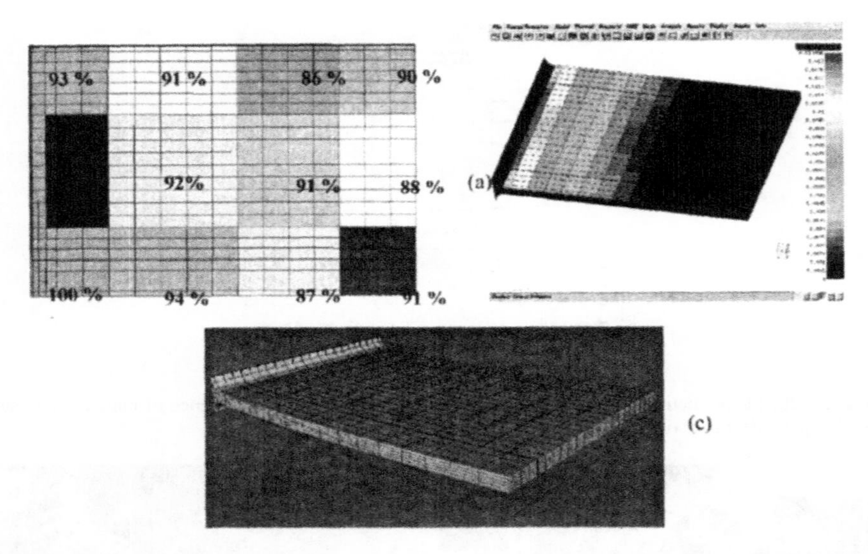

Figure A23: (a) Coating thickness variations as measured and included in the FE models (b) contribution to added damping (Predictions- Mode 1F static) (c) modeshape of the fundamental vibration mode.

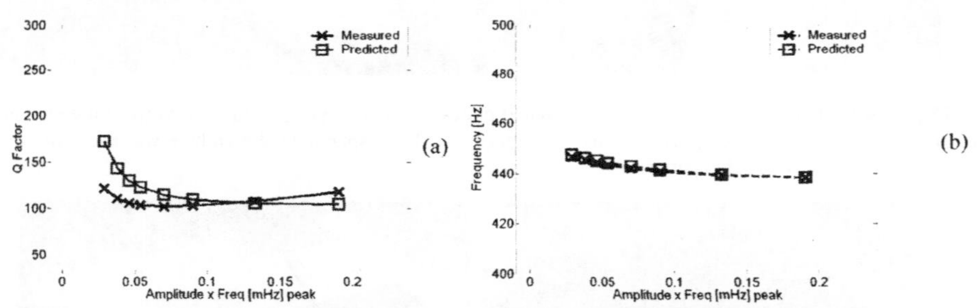

Figure A24: Comparison of measured against predicted results for Mode 1F under static conditions: (a) Q Factors, (b) frequencies.

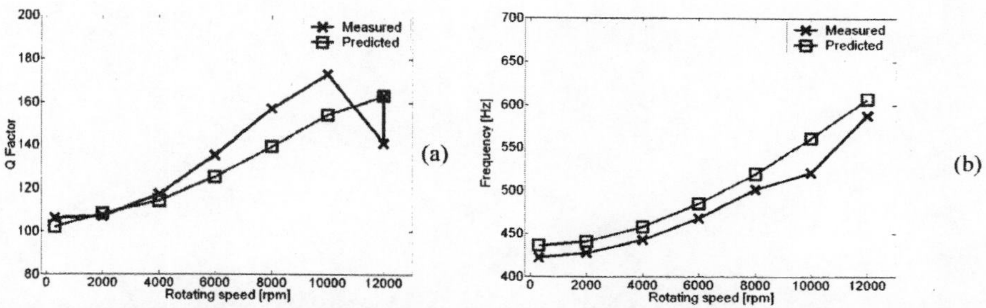

Figure A25: Comparison of measured against predicted results for Mode 1F under rotating conditions: (a) Q Factors, (b) frequencies.

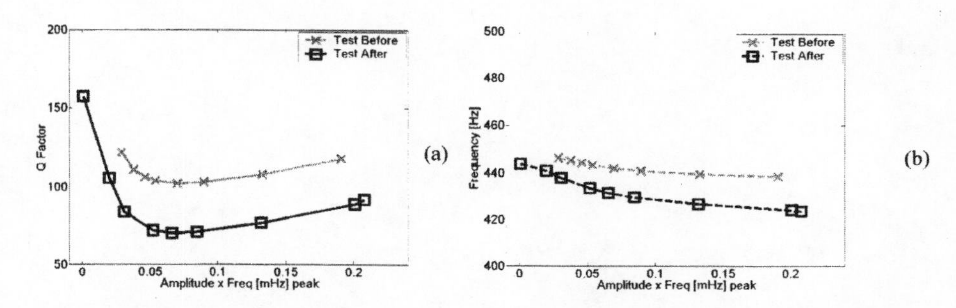

Figure A26: Comparison of measured results under static conditions before and after the rotating tests.

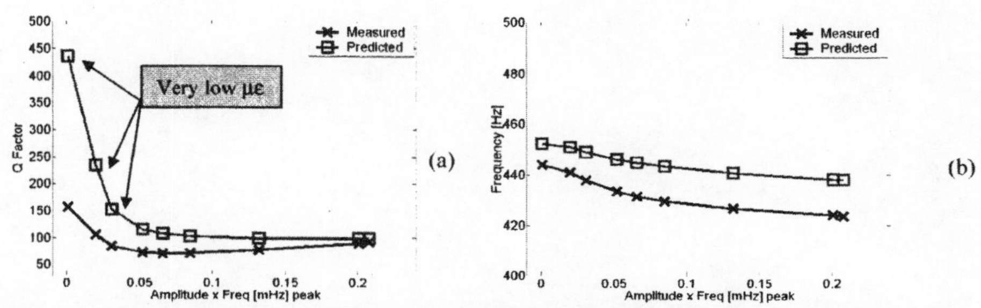

Figure A27: Comparison of measured against predicted results for Mode 1F under static conditions after the rotating tests: (a) Q Factors, (b) frequencies.

DETERIORATION AND RETENTION OF COATED TURBOMACHINERY BLADING

Widen Tabakoff, Awatef A. Hamed, and Rohan Swar
University of Cincinnati
Cincinnati, Ohio, USA

ABSTRACT

The design and development of high-performance turbomachinery operating in an environment with solid particles requires a thorough knowledge of the fundamental phenomena associated with solid particulate flows. Because of the serious consequences of turbomachinery erosion on engine performance and life, it is important to have reliable methods for predicting erosion when solid particles are ingested with the turbomachinery flow. Studies dealing with predicting blade surface erosion intensity and pattern, have been conducted for various blade and coating materials at the University of Cincinnati's Turbomachinery Erosion Laboratory over the past thirty years. This investigation was performed to test new Thermal Barrier coatings for Turbine blades operating at high velocities and temperatures. INCO718 coupons 1x1x1/16" were coated with MDC-150L.Pt-AL and 5 and 10 mils nominal of 7YSZ EB PVD thermal barrier coating (TBC). The samples were tested the erosion rates of the substrate material and TBC at different temperatures, particle impact velocities and impingement angles. Tests were also conducted to study the life of TBC under the same impact conditions.

INTRODUCTION

Suspended solid particles are often encountered in turbomachinery operating environment because of several mechanisms that contribute to particle ingestion in gas turbine engines. Solid and molten particles can be produced during the combustion process from burning heavy oils or synthetic fuels, and aircraft engines can encounter particles transported by sand storms to several thousand feet altitude [1]. Thrust reverser efflux at low airplane speeds as well as engine inlet to ground vortex during high power setting with the aircraft standing or moving on the runway can blow sand, dust, ice, and other particles into the engine. Helicopter engines are especially susceptible to large amounts of dust and sand ingestion during hover, takeoff, and landing. Dust erosion proved so severe during the Vietnam field operations that some engines had to be removed from service after fewer than 100 hours of operations. Particulate clouds from the eruption of volcanoes present one of the most dangerous environments for aircraft engines. Several incidents have been related to engine operation in volcanic ash cloud environments. After two decades of technological advances, the loss in power and surge margins as a result of compressor blade erosion caused some helicopter units to be removed after fewer than 20 hours during the Gulf War field operations. A picture of T-53 compressor after erosion tests conducted with runway sand at the University of Cincinnati is shown in Fig. 1.The light colors indicate the eroded rotor and stator blade surfaces. One can see the leading edge and pressure surface erosion along the first rotor and towards the rotor blade tips and stator blade roots in subsequent stages.

New blade coatings and materials are continuously being developed to meet the challenging requirements of modern gas turbine engines. Because of the serious consequences of blade degradation on gas turbine life and performance, it is necessary to control surface erosion by suspended solid particles that are often encountered in turbomachinery flows. This requires systematic characterization and understanding of the erosion mechanisms that cause coating and blade material erosion. A complex phenomenon such as blade surface deterioration by erosion requires a combination of experimental and computational research efforts [1]. Experimental studies require special high temperature erosion wind tunnels to simulate the wide range of aerodynamic and thermal conditions in modern gas turbines. Erosion test results for gas turbine super alloys and coatings demonstrated that the eroding particle characteristics are affected by temperature and impact conditions. [2-4]. Particle size also affects the blade impact patterns since smaller particles tend to follow the flow while larger particles impact the vane and rotor blades. However, even particles of one to thirty microns have been known to damage exposed components of coal-burning turbines [14]. Numerical simulations of the particle trajectories through gas turbine engines provide valuable information on the vane and blade impact patterns [1, 15, 16, 17, 18], but accurate correlations based on reliable measurements of particle restitution characteristics in tunnels equipped with optical access [19, 20].

In the current work, an experimental investigation was conducted to study blade and coating material erosion by particle impacts under conditions that simulate those encountered in modern gas turbines. The tests were conducted in the hot erosion facility over a range of impacting particle velocities, and high test temperatures impact angles. Results are presented for the experimentally measured erosion rates under these conditions for both coated and uncoated blade materials. The results demonstrate the effect erosion by particle impacts on the life of thermal barrier coating.

EROSION WIND TUNNEL AND TEST PROCEDURE

The University of Cincinnati (UC) erosion wind tunnel facility is shown schematically in Fig. 2. It consists of the following components: particle feeder (A), main air supply pipe (B), combustor (C), particle pre-heater (D), particle injector (E), acceleration tunnel (F), test section (G), and exhaust tank (H). Abrasive particles of a given constituency and measured weight are placed into the particle feeder (A). The particles are fed into a secondary air source and blown into the particle preheater (D), and then to the injector (E), where they mix with the primary air supply (B), which is heated by the combustor (C). The particles are then accelerated via high velocity air in a constant-area steam-cooled duct (F) and impact the specimen in the test section (G). The particulate flow is then mixed with coolant and directed to the exhaust tank.

Varying the tunnel airflow controls particle velocity, while the particle impingement angle is controlled through the target sample rotation relative to the airflow. Heating the flow stream, which, in turn, heats the erosive media and sample(s), varies the temperature. As can be seen from Figure 1, the tunnel geometry is uninterrupted from the acceleration tunnel throughout the test section in order to preserve the aerodynamics of the flow passing over the sample(s).

The tests were carried out for coupons of coated and uncoated turbine blade materials. The coupons were mounted on a sample holder and placed in the erosion wind tunnel at the designated angles and subjected to erosion by a calibrated mass of particles. The holder protected all but one target coupon surface that was exposed to particle impacts. The samples were weighed and their surface roughness was measured using a Taylor Hobson Talysurf before and after the erosion tests. Post erosion surface traverses were centered on the eroded portion of the sample.

The tested samples are evaluated before and after they are tested in the tunnel with a calibrated mass of erosive particles to determine the eroded weight by the impacting particles. The erosion rate, ε measured in mg/g is defined as the ratio of the measured coated sample erosion mass loss ΔW to the mass of erosive impacting particles Q_p.

$$Erosion\ rate = \frac{change\ in\ mass\ of\ sample}{mass\ of\ impacting\ particles}$$

$$= \varepsilon = \frac{mg}{g} = \frac{\Delta W}{Q_p}$$

The erosion rate prediction can be obtained by subtracting the tested sample weight after erosion from the initial sample weight to obtain ΔW = weight loss. To convert the weight loss ΔW in volume loss ΔV, divide ΔW by the density (ρ) of the coating on the tested material.

$$\varepsilon_v = \frac{\Delta V}{Q_p} = cm^3/g$$

or

$$\varepsilon_w = \frac{\Delta W}{Q_p} = mg/g$$

RESULTS AND DISCUSSIONS

Two types of erosion tests were conducted first to obtain thermal barrier coating and substrate material erosion rates then to measure the coating life. Test samples of 1"x1"x1/16"" INCO718 coupons coated with 10 and 5 mils nominal of 7YSZ EB PVD thermal barrier coating were used in the erosion rate and TBC erosion life tests respectively. Experimental measurements were obtained for 26 micron Aluminum Oxide particles at of 400, 800, and 1,200 ft/sec impact velocities, and at 1600 °F, 1,800 °F, and 2,000 °F test temperatures. Test data were accumulated by setting the particle impingement angles at 20°, 45°, 70°, and 90°.

Fig. 3 presents the experimentally measured erosion rates variation with particle impingement angles for INCO 718 substrate and for 10 mil TBC coated sample at temperature 2,000 °F, Velocity 1,200 ft/s and sample impact particles Q_p = 5 grams. Inspection of this figure shows typical erosion behavior of ductile and non-ductile materials for the substrate and coating respectively. The erosion rates angles for ductile material is maximum between 20° – 30° and for non-ductile materials is maximum at 90°.

Fig. 4 presents the experimentally measured erosion rate variation with particle impingement angles for the 10 mil TBC coated samples at 1800 °F and particle impact velocities of 400, 800, and 1,200 ft/sec. According to these results the erosion behavior is typical of that for non ductile material with maximum erosion rate at 90°. As expected, the erosion rate increases with particle impact velocity at all impingement angles. Fig. 5 presents the experimentally measured erosion rate variation with particle impingement angles for the 10 mil TBC coated samples at 1,600 °F and particle impact velocities of 800 ft/s and 1,200 ft/sec. Comparing the two figures one can observe the increase in erosion rate with test temperature and velocity.

A second set of tests were conducted to evaluate the thermal barrier coating life periods at different particle impact conditions. In these tests the INCO718 samples, coated with 5 mil TBC were tested with different erosive particle masses of Q_p = 2.5 grams and 5 grams for comparison. The erosion rate results obtained with Q_p = 2.5 grams were comparable to those in Fig. 4. On the other hand the results for the measured erosion rate with Q_p = 5 grams were different as presented in Fig. 6 & 7. The measured erosion rates at impact angles up to 60° were comparable to those obtained for 10 mil TBC coating with Q_p = 5gm. However, the measured erosion rate at 90° was lower than that for the 10 mil TBC coating, but higher than substrate erosion rate at the same impingement angle (Fig. 6). This indicates that above 60° impingement angle 5 mils TBC was completely removed with further erosion of the substrate leading to the reduced erosion rate above 60° impingement angle. This was consistent over the range of tested velocities as shown in Fig. 7. According to these results careful consideration must be given to the TBC thickness to ensure its life under impact by erosive particles.

SUMMARY

The presented erosion results indicate that the erosion rate variation with particle impingement angles is very different for TBC and substrate blade material. TBC erosion increases with impingement angles while maximum substrate erosion rate is observed between 20° and 30°. The erosion rate of TBC increases with particle impact velocity and with test temperatures. Increased erosive particle masses were used to study TBC erosion life. The conclusion is that the present 10 mils TBC coating provides good erosion protection over the range of tested velocities, temperatures and impact angles.

ACKNOWLEDGEMENT

This research work was supported by NASA contract NNX07AC69A, Project Manager Dr. Robert Miller.

REFERENCES

1. Hamed, A. & Tabakoff, W., "Experimental and Numerical Simulation of Ingested Particles in Gas Turbine Engines," AGARD (NATO) 83rd Symposium of Propulsion and Energetics Panel on Turbines, Rotterdam, The Netherlands, 25-28 April 1994.
2. Tabakoff, W., "High-temperature erosion resistance coatings for use in turbomachinery," Wear, Vols. 186-187, 1995: pp. 224-229.
3. Tabakoff, W., "Protection of coated superalloys from erosion in turbomachinery and other systems exposed to particulate flows," Wear, Vols. 233-235, 1999: pp. 200-208.
4. Tabakoff, W., Hamed, A., & Wenglarz, R., "Particulate Flows, Turbomachinery Erosion and Performance Deterioration," *Von Karman Lecture Series* 1988-89, May 24-27, 1988, Brussels, Belgium.
5. Balan, C. & Tabakoff, W., "Axial Flow Compressor Performance Deterioration," AIAA 84-2108, 1984.
6. Hamed, A., Tabakoff, W. & Singh, D., "Modeling of Compressor Performance Deterioration Due to Erosion," International Journal of Rotating Machinery, Vol. 4, November 1998, pp. 243-248.
7. Taylor, R.P., "Surface Roughness Measurements on Gas Turbine Blades," ASME Journal of Turbomachinery, Vol. 112, 1990: pp. 175-180.
8. Tarada, F. & Suzuki, M., "External Heat Transfer Enhancement to Turbine Blading Due to Surface Roughness," ASME Paper 93-GT-74, ASME IGTI, Cincinnati, Ohio, May 1993.
9. Bons, J.P., Taylor, R.P., McClain, S.T., & Rivir, R.B., "The many faces of turbine surface roughness," 2001-GT-0163, Proceedings of ASME Turbo Expo, June 2001, New Orleans, LA.
10. Blair, M.F., "An Experimental Study of Heat Transfer in a Large-Scale Turbine Rotor Passage," ASME Journal of Turbomachinery, Vol. 116, 1994: pp. 1-13.
11. Hoffs, A., Drost, U., & Bolcs, A., "Heat Transfer Measurements on a Turbine Airfoil at Various Reynolds Numbers and Turbulence Intensities Including Effects of Surface Roughness," ASME Paper 96-GT-169, IGTI, Birmingham, UK, June 1996.
12. Bogard, D.G., Schmidt, D.L., & Tabbita, M., "Characterization and Laboratory Simulation of Turbine Airfoil Surface Roughness and Associated Heat Transfer," ASME Journal of Turbomachinery, Vol. 120, 1998: pp. 337-342.

13. Abuaf, N., Bunker, R.S., & Lee, C.P., "Effects of Surface Roughness on Heat Transfer and Aerodynamic Performance of Turbine Airfoils," ASME Journal of Turbomachinery, Vol. 120, 1998: pp. 522-529.
14. McCay, L., "The Coal Burning Gas Turbine Project," Report of Interdepartmental Gas Turbine Steering Committee, Australian Government Publishing Service, 1973.
15. Tabakoff, W., Hamed, A. & Metwally, M., "Effect of Particle Size Distribution on Particle Dynamics and Blade Erosion in Axial Flow Turbines," Journal of Gas Turbine and Power, Vol. 113, October 1991, pp. 607-615.
16. Metwally, M., Tabakoff, W. & Hamed, A., "Blade Erosion in Automotive Gas Turbine Engine," Journal of Engineering for Gas Turbines and Power, January 1995, Vol. 117, pp. 213-219.
17. Hamed, A. & Kuhn, T.P., "Effects of Variational Particle Restitution Characteristics on Turbomachinery Erosion," Journal of Engineering for Gas Turbines and Power, July 1995, Vol. 117, pp. 432-440.
18. Hamed, A., "Influence of Secondary Flow on Turbine Erosion," Journal of Turbomachinery, Vol. 111, No. 3, pp. 310-314, July 1989.
19. Tabakoff, W., Malak, M.F. & Hamed, A., "Laser Measurements of Solid Particles Rebound Parameters Impinging 2024 Aluminum and 6A1-4V Titanium Alloys," AIAA Journal , Vol. 25, No. 5, pp. 721-726, May 1987.
20. Tabakoff, W., Murugan, D.M. & Hamed, A., "Effects of Target Materials on the Particle Restitution Characteristics for Turbomachinery Application," AIAA Paper No. 94-0143, January 1994.

Fig.1 Multistage compressor erosion

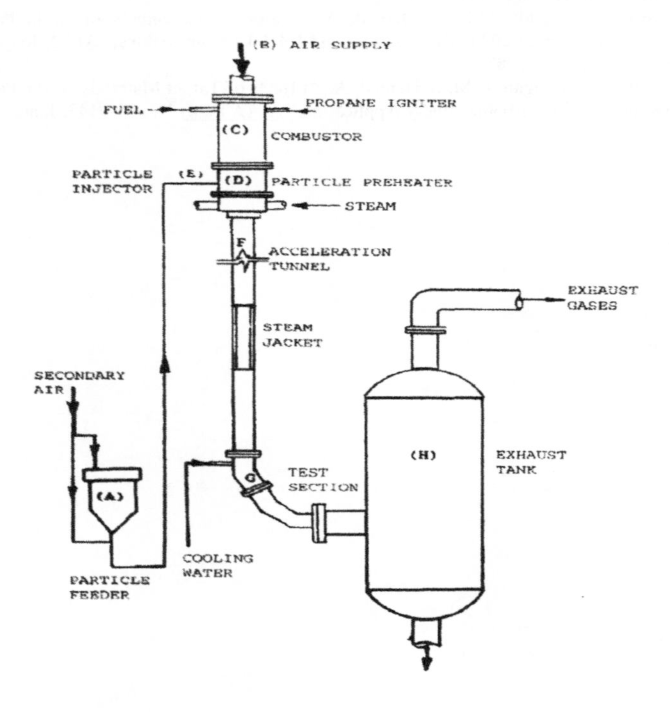

Fig.2 Schematic of Erosion Test Facility

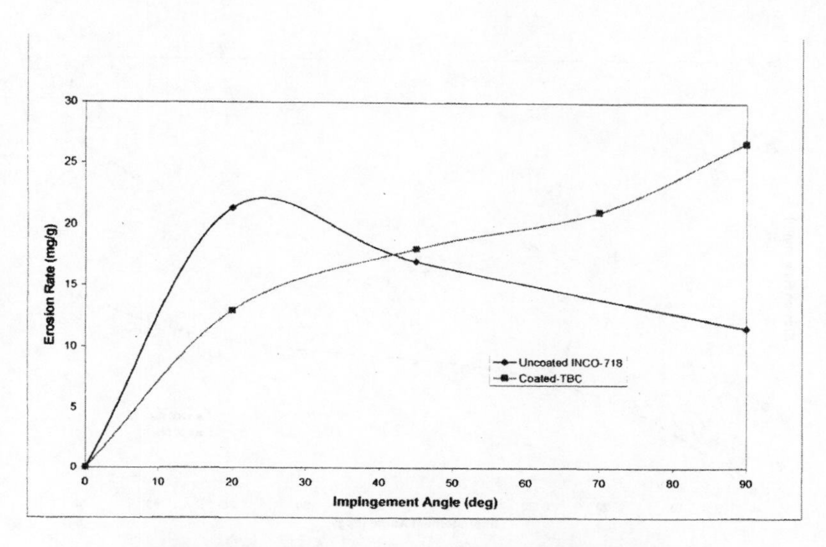

Fig.3 Measured erosion rates of coated and uncoated samples (10 mils coating, T = 2,000° F, V= 1,200 ft/s, Q_p = 5 gm)

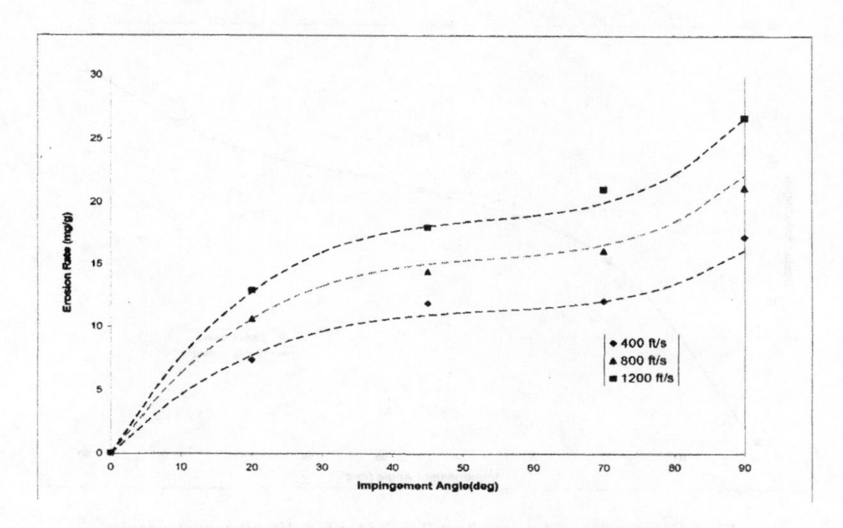

Fig. 4 TBC Erosion Test Results (10 mils coating, T = 1,800° F, Q_p = 5 gm)

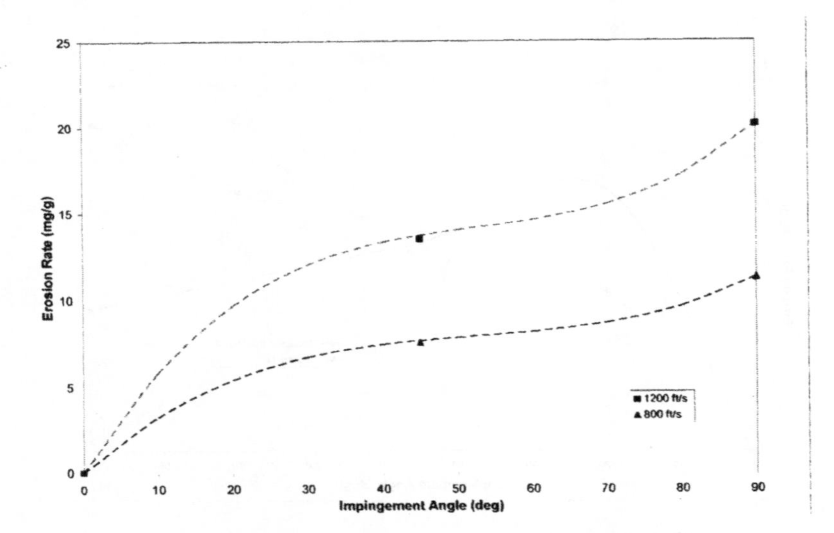

Fig.5 TBC Erosion Test Results (10 mils coating, T = 1,600° F, Q_p = 5gm)

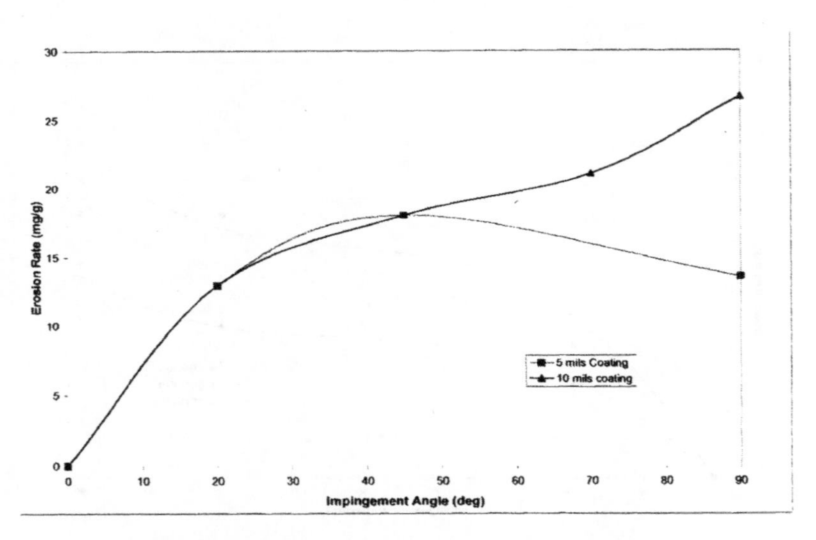

Fig.6 Comparison of Erosion Test Results for 5 & 10 mil coated samples (T = 2,000 °F, V = 1,200 ft/s, Q_p = 5 gm)

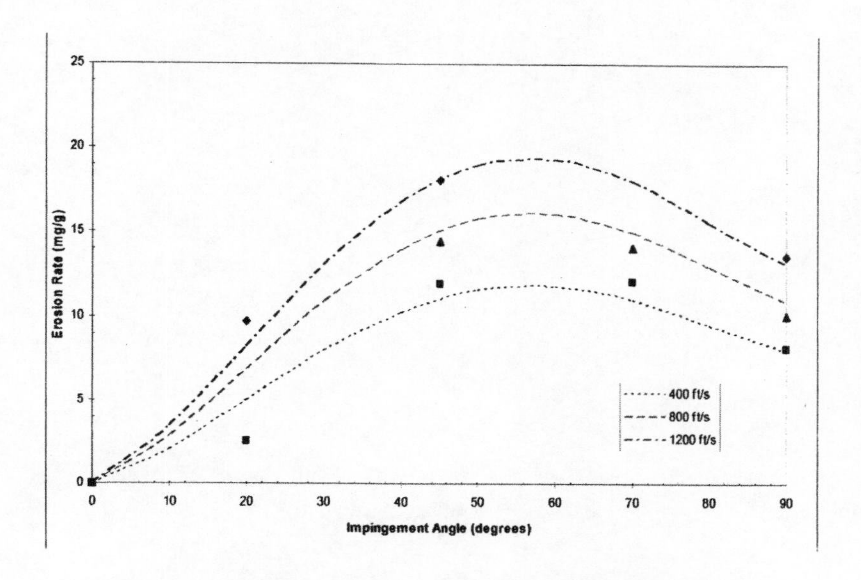

Fig. 7 5 mil TBC Erosion Life Test Results (T = 2,000 °F, Q_P = 5 gm)

LARGE AREA FILTERED ARC AND HYBRID COATING DEPOSITION TECHNOLOGIES FOR EROSION AND CORROSION PROTECTION OF AIRCRAFT COMPONENTS

V. Gorokhovsky[a], J. Wallace[a], C. Bowman[a], P.E. Gannon[a], J. O'Keefe[a], V. Champagne[b] and M. Pepi[b]
[a]Arcomac Surface Engineering, LLC – Bozeman, MT USA
[b]US Army Research Laboratory – Aberdeen Proving Ground, MD USA

ABSTRACT

Erosion and foreign object damage are among the most frequent failure modes for high speed rotating components of turbine engines such as compressor blades and helicopter rotor blades. Environmental corrosion is also known to accelerate degradation. Large Area Filtered Arc Deposition (LAFAD) technology and hybrid filtered arc-assisted electron beam physical vapor deposition (EBPVD) or unbalanced magnetron (UBM) sputtering processes were evaluated for deposition of hard cermet coatings to improve erosion resistance performance over a wide range of impact angles, while simultaneously maintaining aerodynamic properties and improving corrosion resistance of coated components. Coating performance was evaluated as a function of substrate bias voltage. Different grades and speeds of erosive particles were used to simulate different operation conditions in compressor blade and helicopter rotor blade applications. Simulated salt fog testing was used to assess corrosion resistance of different coatings on substrates made of high chromium bearing steel. Increased component performance can be attributed to the multi-layered nanocomposite morphology of LAFAD coatings with low defect density, which effectively encapsulates the substrate surface with a nanolaminated metal-ceramic shield, and also impedes perpendicular crack propagation. In addition, hybrid LAFAD+EBPVD or LAFAD+UBM cermet coatings have substantially-improved coating density compared to conventional EBPVD or UBM coatings. In this work, characterization of erosion and environmental corrosion resistance performance of several LAFAD and hybrid LAFAD+EBPVD and LAFAD+UBM nanostructured coatings will be presented, with possible applications discussed.

INTRODUCTION

Gas turbine engines operating in arid environments suffer significant wear from ingested sand that causes erosion damage to various components[3,4-6]. Damage caused by solid particle impact reduces the engine life as well as engine power and efficiency, and results in high engine maintenance costs[5]. Several turbine engine components operate at elevated temperature, which accelerates erosion/corrosion degradation. In this case both particle impact wear and forming of oxide scale at elevated temperature can contribute to degradation of these components[2,17]. This problem is exacerbated in the case of blisks and impellers where the disc and the associated blades are all manufactured as a single piece; hence, damaged blades cannot simply be removed and replaced[5]. In case of helicopter rotor blades, the size and characteristic impact velocity of the particulate flow are several times larger than that of rotating components of turbine engine due to the absence of any protective filters, which are now installed by most turbine engine manufacturers[7]. In addition, helicopter blades often suffer severe damage from rain erosion.

Erosion-protective surface treatments consist of different metal, ceramic or cermet coatings deposited by thermal spray, CVD or PVD techniques[4,6]. The most popular are TiN-based coatings deposited by EBPVD, as well as magnetron sputtering and cathodic arc evaporation techniques[3-6,10,11]. Among PVD and CVD coatings, the most promising results have been realized using TiN, ZrN, TiCrN and TiAlN compositions, both monolithic and with multilayer architectures (using a sequence of metallic and ceramic sublayers). However, these coatings have columnar morphology with relatively large column size, which increases the surface roughness and accelerates degradation due to both development of corrosion pits and high temperature oxidation[2,17]. Hard coatings are inherently brittle and are prone to cracking and spallation under high energy and near 90° impacts; this produces the paradoxical result that an increase in hardness can result in a decrease in wear resistance. On the other hand, tough materials are prone to cutting and micromachining wear mechanisms at low angles of impact. The solution is to combine the hard impact resistance of ceramics with the toughness of metals by forming multilayer erosion-resistant coatings[10,11]. Recently, nanostructured composite coatings with superior functional properties have been developed[8,9,16,18]. They utilize nano-laminated or superlattice coating architectures or use a ceramic nanopowder as a filler in a polymer or metal matrix as in the case of nano-laminated Si_xC_y/DLC and polymer-ceramic coatings recently tested as erosion resistant coatings for V-22 aircraft application[12,13]. To meet these requirements, there is growing interest in using highly ionized metal vapor for deposition of dense cermet coatings with smooth surfaces, reduced defect density, and fine grain morphology. Arcomac's Large Area Filtered Arc Deposition (LAFAD) process is capable of producing gaseous and metal-gaseous plasmas with a high ionization rate reaching 100% for metal vapor and more than 50% for the gaseous plasma component with negligibly low concentration of macroparticles[14]. Arcomac's recently developed Filtered Arc Plasma Source Ion Deposition (FAPSID) surface engineering process utilizes LAFAD in hybrid combination with conventional PVD (unbalanaced magnetrons (UBM), EBPVD, thermal evaporators) and low pressure CVD vapor plasma sources in one processing chamber as schematically illustrated in Figure 1[15,16]. In this work, the LAFAD technology and hybrid filtered arc assisted EBPVD and magnetron sputtering processes were evaluated for deposition of hard cermet coatings to improve erosion resistance performance, while simultaneously maintaining aerodynamic properties and improving corrosion resistance of coated components. Increased component performance can be attributed to the defect-free multi-layered nanocomposite morphology of LAFAD coatings, which effectively encapsulates the substrate surface with a nanolaminated metal-ceramic shield and prevents perpendicular crack propagation.

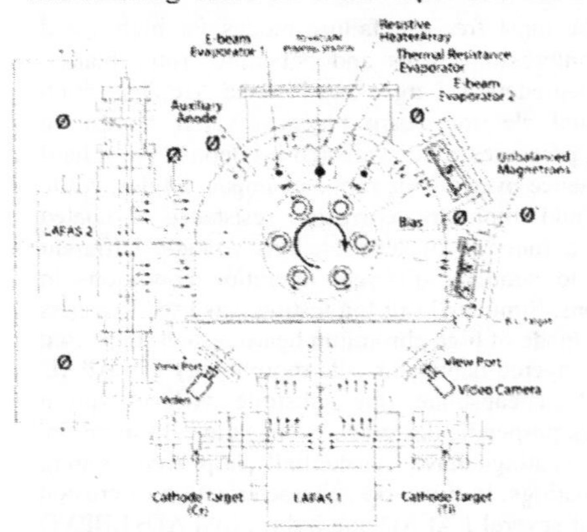

Figure 1. Schematic illustration of FAPSID surface engineering system.

EXPERIMENTAL DETAILS

Coating Deposition and Basic Properties Characterization

In this work, the FAPSID surface engineering system was used for producing various filtered arc and hybrid coatings by deposition from highly ionized metal-gaseous vapor plasmas. First, LAFAD multilayer TiN/Ti coatings were deposited on substrates made of stainless steels and of Ti6Al4V alloy. In several deposition runs, a 50V, 100V and 150V bias potential were applied to the substrates, while other parameters remained constant. The thickness of the coatings produced in these runs ranged from 3 to 15 μm. In a subsequent process, dual segment coatings were produced by hybrid LAFAD+UBM technology. In this coating design, the bottom segment interfacing the steel substrate was made of TiCrN/TiCr multilayer (6 or 12 μm) followed by a BCN top segment coating (2μm). A cross-sectional electron micrograph of this coating is shown in Figure 2. Finally, several oxiceramic multilayer nanostructured coatings of YSZ+$(Al,Cr)_2O_3$ composition, having thicknesses ranging from 30 to 100μm, were produced using filtered arc ionized EBPVD process.

The basic coating mechanical properties were characterized by hardness, adhesion, residual stress and surface profile. Coating thickness was determined by the CALOtest™ spherical abrasion technique and optical micrometry to an accuracy of +/-0.1μm. Coating composition was characterized by Rutherford backscattering spectroscopy (RBS) and energy dispersive x-ray spectroscopy (EDS). Scanning electron microscopy (SEM) was used for surface morphology assessments, cross section imaging and elemental analysis/mapping (with EDS). Coating hardness and Young's modulus were measured by a MTS XP nanoindenter with a CSM module and Berkovich tip.

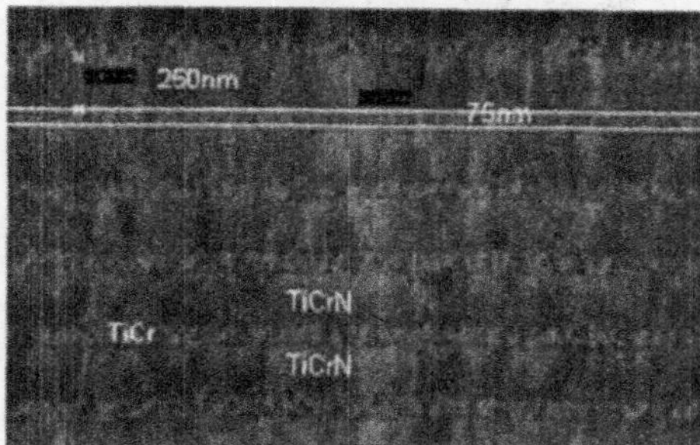

Figure 2. SEM cross-section image of a TiCr/TiCrN multilayer bottom segment coating.

Table I. Basic Coating Properties.

Coating Property	TiCr/ TiCrN	Ti/ TiN (LAFAD)	Ti/TiN (LAFAD)	Ti/TiN (LAFAD)	Ti/TiN (LAFAD)	BCN (LAFAD+ UBM)	TiN monolithic (EBPVD)
Substrate	Pyrowear 675	440A	E/formed Ni	AM355 steel	Ti6Al4V	Pyrowear 675+TiCrN coating	Ti6Al4V
Thickness. μm	6; 12	3-12	15	15	15	2	12
Hardness. GPa	20	25 (50V) 30 (100V) 23 (150V)	25 (50V) 30 (100V) 23 (150V)	25 (50V) 30 (100V) 23 (150V)	25 (50V) 30 (100V) 23 (150V)	43	25
Residual stress, GPa	0.3 tensile	<1 compres.	<1 compres.	<1 compres.	<1 compres.	3.0 compres.	3-4 compres.
Elastic Modulus, GPa (bias voltage)	305	397 (50V) 420 (100V) 280 (150V	397 (50V) 420 (100V) 280 (150V)	397 (50V) 420 (100V) 280 (150V	397 (50V) 420 (100V) 280 (150V	415	
H/E Ratio	0.07	0.066 (50V) 0.071 (100V) 0.08 (150V)	0.066 (50V) 0.071 (100V) 0.08 (150V)	0.066 (50V) 0.071 (100V) 0.08 (150V)	0.066 (50V) 0.071 (100V) 0.08 (150V)	0.1	
Composition, at%	Ti: 49.6 Cr: 12.9 N: 37.9	Ti: 53.9 N: 46.1	Ti: 53.9 N: 46.1	Ti: 53.9 N: 46.1	Ti: 53.9 N: 46.1	B: 77.6; C: 14.4; N: 2.1; O: 2.1; Ar: 1.0	Ti: 52.1 N: 47.9
Critical Load, N (Lc2/Lc1)	Lc1=35 Lc2=175	Lc2=34					Lc2=25
Rockwell adhesion[19]	HF1	HF1	HF4	HF1	HF1	HF2	HF2

Effective Young's modulus $E^*=E/(1-n^2)$, where n is Poisson's ratio of 0.2, and resistance to plastic deformation (H^3/E^{*2}) ws calculated from obtained hardness (H) and Young's modulus (E) data as in[18]. Coating adhesion was assessed by the standard Rockwell indentation method using a 120° diamond indenter and a 1500N load[19]. CSEM Revetest equipment was used for quantitative scratch test assessment of the coating adhesion/cohesion with the following test parameters: linear loading from 0-

200N, 100N/min loading rate, 6.5mm/min travel rate, 0.2mm radius diamond tip. Coating failure was assessed by acoustic emission and tangential force monitoring during testing, and was confirmed by post-test back-scatter electron imaging (BEI) and secondary electron imaging (SEI). Surface profiling and mapping were performed using a Veeco Dektak 8 contact profilometer (5μm conical tip, vertical resolution ~10nm, x-resolution ~100nm, y-resolution ~3000nm). Typical basic coating properties are presented in Table I.

Examination of Eroded Surfaces

The investigation of a single impact area was used during a preliminary study aimed at better understanding of coating degradation in separate collisions. Preliminary erosion tests were conducted on 301 stainless steel coupons with a 10 μm thick TiN/Ti multilayer coating, using conventional sandblasting equipment (Pangborn) with a fixed nozzle at different air tank pressures. The characteristic parameters of the sandblasting test were the following: erodent media was silica 230mesh/74μm; air tank pressure ranged from 20 to 80PSI (corresponding to average velocity ranging from 100 to 150m/s); impact angle was 90°; and exposure time was 5-15s. Optical reflective microscopy and SEM were used for assessment of the erosion area. Both SEI and BEI were employed for topological and compositional images of the erosion area. In the BEI mode, lighter elements (e.g., Ti and N) appear darker than the heavier elements (e.g. Fe, Cr, Ni). As a result, islands of exposed steel surface appear as white spots surrounded by darker TiN coated areas. Analysis was conducted at isolated, single impact spots located at the periphery of the central erosion zone created by five seconds of sandblasting at different pressures, which allowed a reduction of the uncertainty related to multiple impact areas located near the center of the erosion zone. The objective of this analysis was to characterize the erosion mechanism vs. impact energy of blasting particles.

Erosion Testing at High Impact Energy

Erosion resistance performance of deposited coatings for potential application on protective shields for helicopter rotor blades requires higher impact energies and larger erodent media than that of turbine engine components. Methodical testing of the coating performance was conducted at the University of Dayton Research Institute (UDRI). The schematic of the UDRI dust rig is shown in Figure 3. A six inch square test area is uniformly covered with a pre-determined mass of particles of a known size at a speed measured by a laser Doppler anemometry system, which determines a delivery pressure for the required velocity. The exact parameters for the erosion test at UDRI used in this work are as follows. Square sample coupons (2"x2"x1/8") of AM355 steel, electroformed nickel (EF Ni - consisting of 75μm Ni electroplated steel) and Ti4Al6V alloys were used as candidate materials for the protective shields of helicopter blades. Sand erodent (from Bagram, Iraq) 177-250μm was delivered to the samples at average velocity of 222.5m/s, for a total erodent loading ranging from 10-26g/cm^2. Note that the commonly used protocol for erosion testing of compressor blades uses a media size (alumina) of ~50-80um, and a nominal velocity ranging from 60 to 80m/s[3,4,6]; this represents 500 to 1000 times less kinetic energy per particle compared to the test at UDRI.

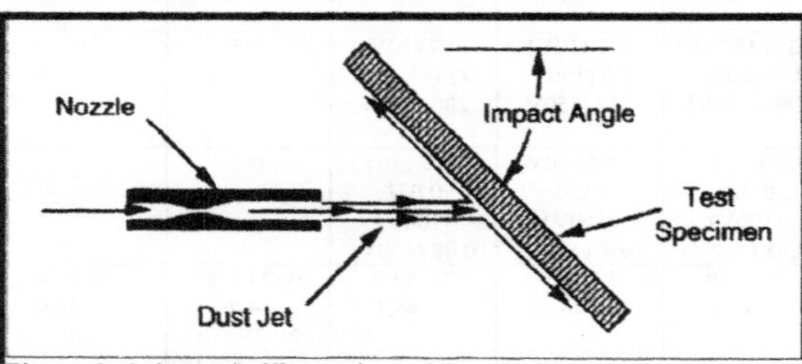

Figure 3. Schematic illustration of the erosion test rig at UDRI.

Erosion Testing at Moderate Impact Energy

Quantitative evaluation of erosion resistant properties was performed at Arcomac Surface Engineering with the use of a FALEX F-1507 Air Jet Erosion Test Rig. The test parameters (based on the ASTM G76 "Test Method for Solid Particle Erosion") were the following; alumina erodent (50μm) fed at 2g/min at both 40°, 60° and 90° incidence.

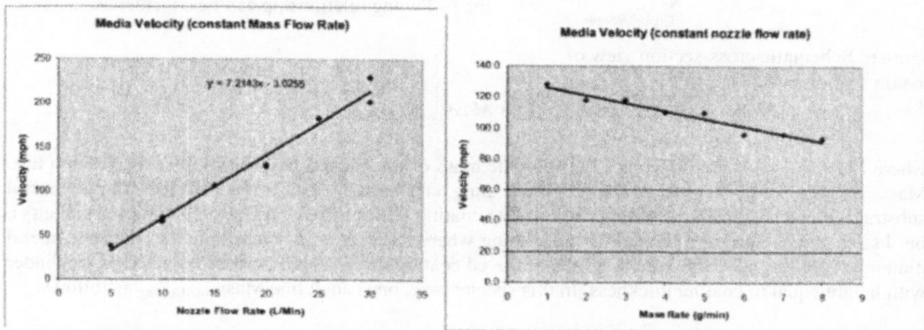

Figure 4. Results of velocity measurement in FALEX F-1507 Air Jet Erosion Test Machine by Dual Disc Method: velocity change vs. nozzle flow rate at constant erodent mass rate (8 g/min) (left); velocity change vs. erodent mass rate at constant nozzle flow rate (15 L/min) (right).

The velocity of the erosion media was measured using the Dual Disk Method first developed by Ruff and Ives[1]. It was determined that two different factors would affect the media velocity: the nozzle flow rate and the mass rate of media. Two separate sets of measurements were taken while holding each of these factors constant. Figure 4 (left) shows the velocity range of the Falex erosion rig while holding the mass rate constant at 8 grams per minute. Figure 4 (right) shows the velocity change over the range of the mass rate of the Falex erosion rig while holding the nozzle flow rate constant at 15 liters per minute. While varying the Mass Rate of erosion media in grams per minute did affect the velocity, the velocity was most stable in the range of 4 – 5 grams per minute of erosion media.

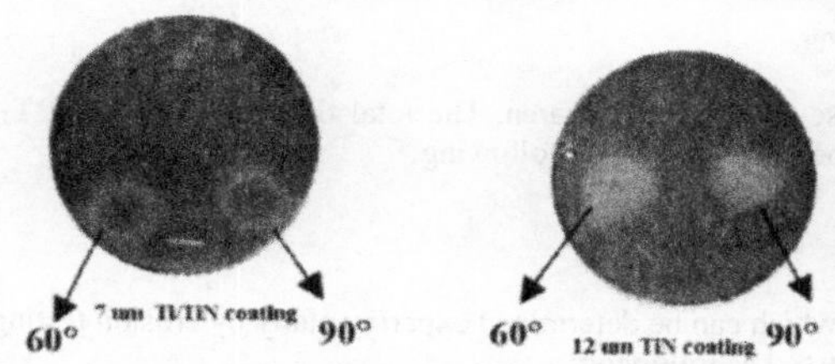

Figure 5. Erosion scars prepared at different impact angles on titanium coupons with 7 μm Ti/TiN multilayer coating and 12 μm TiN monolithic coating.

In the case of relatively thin coatings, the coating exposure occurs at the very beginning of the test run as illustrated from the photographs of erosion area on coupons with 7μm and 12μm thick TiN coatings (Figure 5). This complicates the experimental evaluation of different coatings. To get a quantitative estimation of erosion rate in the case of very thin coatings, the following methodology was developed: in order to find the erosion rate of the coating, it would appear necessary to find an erosion time that just barely blasts through the coating. This short time has been difficult to prove via experiment, as all tests of thin coatings have blasted through the coating.

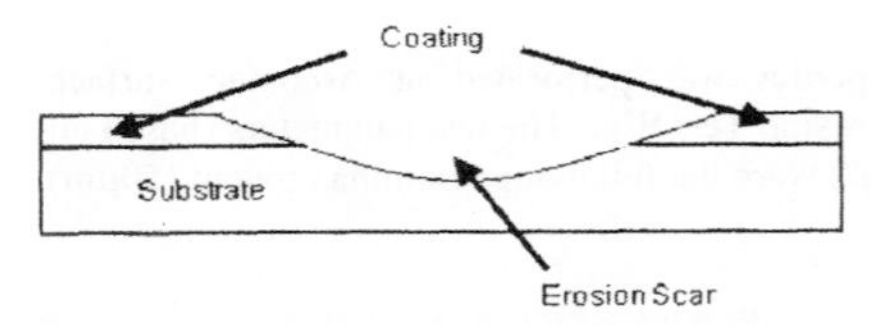

Figure 6. Schematic cross-section view of erosion scar on coated substrate.

The proposed methodology is based on analyzing the weight loss of coated samples when the substrate is exposed by creating a large erosion scar that goes through both the coating and into the substrate material as illustrated in Figure 6. In this geometry the following relationship can be considered:

$$Mass_{Loss-Total} = Mass_{Loss-Coating} + Mass_{Loss-Metal} \quad (1)$$

where $Mass_{Loss-Total}$ is the difference between the mass of the coupon before and after the erosion test, $Mass_{Loss-Coating}$ is the mass loss of the ceramic coating, and $Mass_{Loss-Metal}$ is the mass lost from the metal substrate. Figure 6 shows the erosion scar in the coating with tapered edges (i.e. a conical geometry), but in the special case of a relatively thin coating where the thickness is much smaller than the lateral dimensions of the scar, the volume of the removed coating material can be approximated as a cylinder with height equal to coating thickness. In this special case, one can define $Mass_{Loss-Coating}$ as follows:

$$Mass_{Loss-Coating} = \pi\frac{D^2}{4}h \times \rho_{Coating} \quad (2)$$

where D is the diameter of the erosion scar, h is the thickness of the coating, and ρ is the density of the coating material. In case when the jet strikes the surface of coated coupon at an acute angle (<90°), the erosion wear scar area will have an oval or elliptical shape, such as the one shown in Figure 5 (7μm, 60°). In this case the expression (2) will change to the following:

$$Mass_{Loss-Coating} = \pi abh\rho_{Coating} \quad (2')$$

where a and b are half axes of the ellipse-shaped erosion area. The total time of erosion test T_{Total}, which is determined experimentally, can be expressed as the following:

$$T_{Total} = t_{ceramic} + t_{metal} \quad (3)$$

Erosion rate of the metal substrate $\dot{e}_{metal}$, which can be determined experimentally by erosion testing of the uncoated metal substrate, can be expressed as following:

$$\dot{e}_{metal} = \frac{Mass_{Loss-Metal}}{t_{metal}} \quad (4)$$

Substituting the values of $Mass_{Loss-Coating}$ from Eqn. (2'), the following expression can be written for the erosion rate of the coating in a general case where the jet strikes the surface at an acute angle:

$$\dot{e}_{coating} = \frac{\pi abh\rho_{coating}}{T_{Total} - (Mass_{Loss-Total} - \pi abh\rho_{coating})/\dot{e}_{metal}} \quad (5)$$

Assuming that the feed-rate of the erodent is constant as a function of time, then T_{Total} can be expressed as the total mass of erodent ($Mass_{erodent}$) divided by the feed-rate (R). In this case, Equation (5) becomes:

$$\dot{e}_{coating} = \frac{\pi abh\rho_{coating}}{(Mass_{Erodent}/R) - (Mass_{Loss-Total} - \pi abh\rho_{coating})/\dot{e}_{metal}} \qquad (5')$$

Equation (5') is independent of time measurements, and is based solely on mass ratios and rates. Note: nevertheless, both for test accuracy and from practical considerations, it is preferable to keep the amount of overblasting as low as possible.

RESULTS AND DISCUSSION

Erosion marks created on the surface of coupon "A" by individual collisions with sandblasting particles are shown in Figure 7 (SEI, right). It can be seen that due to the particle impact, the coating is removed from the surface layer-by-layer as a result of brittle cracks propagating parallel to the coupon surface. The compositional image created by BEI shows no substrate exposure over the entire impact site. In the compositional image (Figure 7, left), the stainless steel substrate appears in very limited areas surrounded by multiple layered coating with surface erosion patterns. In all images presented in Figure 7, several layers of multilayer coating are visibly exposed in each single impact erosion site. It was also found that erosion resistance was significantly different for different coating thicknesses and different particulate impact velocities (based upon carrying gas pressure). In addition, the characteristic diameter of single impact scars taken from different spots located at the periphery of the erosion area increases when sandblasting process pressure increases.

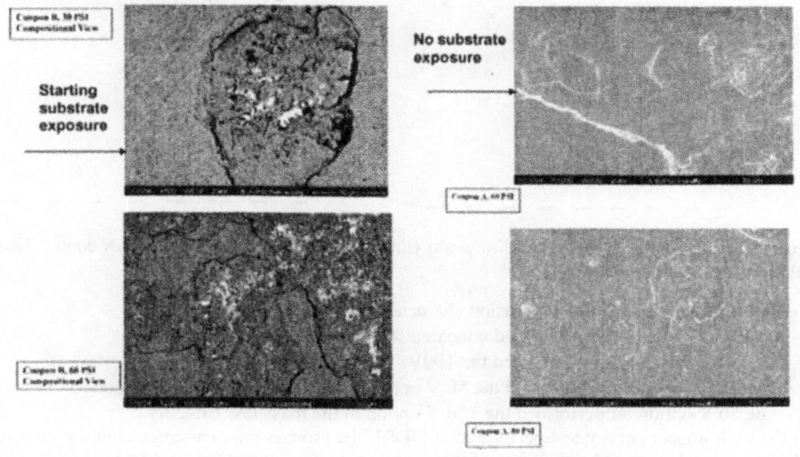

Figure 7. Single impact erosion scars obtained by SEI (right) and BEI (left).

Qualitatively similar results were collected using FALEX test rig. These preliminary erosion tests were performed at two impact angles: 90° and 60°. All tested TiN/Ti multilayer coatings had a thickness of 15 +/-1 μm. The substrate material was either 300-series stainless steel or Ti6Al4V alloy. Three groups of samples were tested: coatings deposited with 50V, 100V and 150V applied bias. The substrates were analyzed by precision surface profilometry and gravimetrical weight loss analysis. Based on mass loss analysis all TiN/Ti multilayer coated samples are outperforming the uncoated steel and titanium samples in terms of erosion wear resistance. Figure 8 shows the surface profile of erosion scar of the uncoated (left) and Ti/TiN multilayer coated Ti6Al4V (right). From the surface profile and volumetric analysis, it is clear that erosion impact inflicts substantially smaller surface deformation on coated vs. uncoated metal samples. It was found that mass loss of uncoated titanium is much less than it can be estimated from the volume of erosion test induced depression. The reason for this volume/mass loss discrepancy is plastic deformation resulting in a pile-up of the Titanium around the wear scar (see Figure 8). A similar screening effect of hard coatings is well known in the case of surface indentation, when the coating functions as a hard membrane, resisting against surface deformation[20].

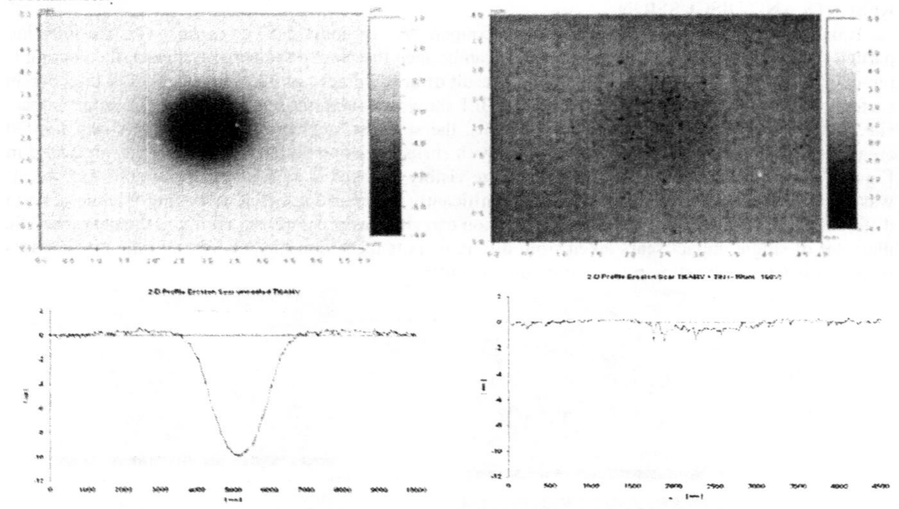

Figure 8. Surface profiles (upper) and line scans (lower) of uncoated (left) and Ti/TiN coated Ti6Al4V (right) after exposure to erosion test.

In brief summary, the following erosion characteristics were discovered:

- All coated samples outperformed uncoated stainless steel.
- The 50V and 150V outperformed the 100V.
- The 150 V option outperformed the 50 V option in the volumetric loss category.
- The 50 V option outperformed the 150 V option in the mass loss category.

In the high impact energy test performed at UDRI, the erosion rate was calculated for each run by dividing the mass loss (mg) by the mass of the erodent (g). The mass of the erodent was calculated by dividing the mass load (g/cm^2) by the impact area (cm^2). Arcomac samples were subsequently compared to the results of the uncoated baseline material. Figure 9 shows the typical results of this test. In conclusion, Arcomac's TiN/Ti multilayer coatings having thickness about 15 μm were able to provide marginally improved protection versus the uncoated titanium and AM355 steel substrates at 30° (50V & 150V bias), and versus the AM355 steel substrates at 90° (50 & 150V bias). Coated electroformed Ni substrate did not show improvement in erosion resistance. It is important to note that the erosion test performed at UDRI used a test protocol with 500 to 1000 times higher average kinetic energy per particle than that of the FALEX 1507 air jet tests.

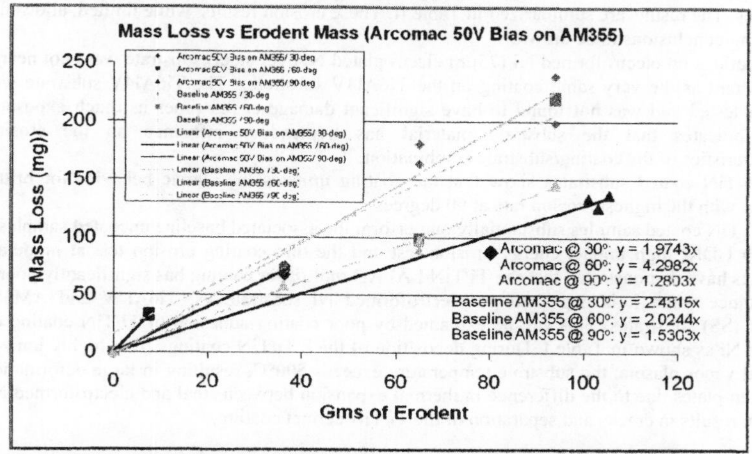

Figure 9. Erosion rate results of Multilayered Ti/TiN (50V Bias) on an AM355 steel substrate compared to uncoated AM355 steel.

The erosion wear of Ti/TiN multilayer cermet coatings deposited on titanium, steel and electroformed Ni substrates was further evaluated under condition of moderate impact energies. Figure 10 presents the relationship between mass loss and the amount of erodent used at 60° and 90° impact angles for Ti6Al4V samples coated with LAFAD multilayer Ti/TiN coatings having 7 and 12μm thicknesses, in addition to 12μm monolithic TiN deposited by a commercial EBPVD process. It can be seen that substrate exposure happens at the very beginning stage of the test for thin coatings (<10μm). The LAFAD Ti/TiN 12μm thick multilayer coating has demonstrated about the same erosion resistance as commercial monolithic TiN coating of the same hardness and thickness.

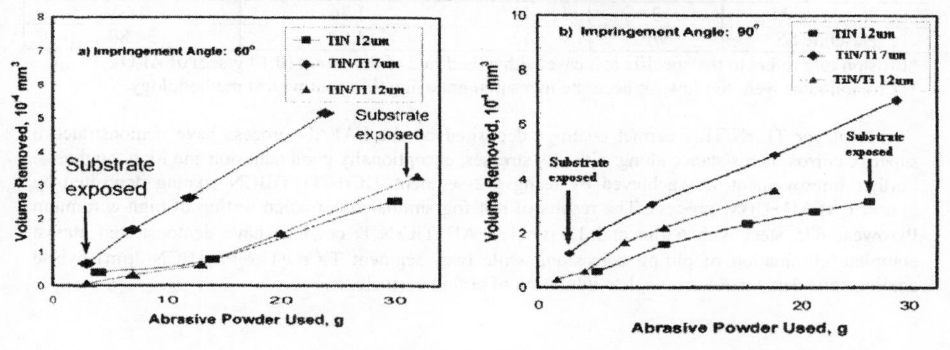

Figure 10. Erosion results of titanium samples with Ti/TiN multilayer coating deposited by LAFAD process vs. conventional monolithic TiN coating (courtesy of Structures, Materials and Propulsion Laboratory, Institute for Aerospace Research, NRC Canada).

The thin coating erosion test methodology outlined above (Eqs. 1-5') has been applied to a series of erosion measurements carried out experimentally using the FALEX erosion-testing rig at moderate impact energies. The results are summarized in Table II. These erosion results, while limited, allowed some interesting conclusions to be drawn:

1. The coatings on electroformed Ni (75μm electroplated Ni over steel) substrate were not nearly as resistant as the very same coating on the Ti6Al4V substrate. The Ti6Al4V substrate was further tested and was not found to have significant damage at 3.5 times as much exposure. This indicates that the substrate material has tremendous influence on the erosion characteristics of the coating/substrate combination.
2. The Ti/TiN coated substrates show that the coating imitates the classic behavior for brittle erosion with the highest erosion rate at 90 degrees.
3. The Ti/TiN coated samples substantially outperform the associated baseline uncoated samples.

Note: both UDRI high impact energy erosion test and the thin coating erosion test at moderate impact energies have demonstrated that the Ti/TiN LAFAD multilayer coating has significantly poorer erosion resistance when it is deposited on electroformed Ni substrate vs. Ti6Al4V and AM355 stainless steel (SS) substrates. This can be explained by poor coating adhesion of Ti/TiN coating on electroformed Ni as shown in Table I. During deposition of thick Ti/TiN coatings from highly ionized LAFAD metal vapor plasma, the substrate temperature exceeds 500°C, resulting in large deformation of the specimen plates due to the difference in thermal expansion between steel and electroformed Ni, which, in turn, results in cracks and separation of the Ti/TiN cermet coating.

Table II. Thin Ti/TiN multilayer coating erosion testing results.

Coupon	Coating Thickness, μm	Impact Angle	Erosion Rate* (μg/sec)	Transfer Rate (μg coating/ grams of erodent)
EF Ni + Ti/TiN	15	90	18.6	131.29
Ti6Al4V + Ti/TiN	15	90	N/A	**
SS + Ti/TiN	3	90	2.18	15.39
Baseline SS	N/A	90	4.04	28.52
SS + Ti/TiN	3	60	.614	4.33
Baseline SS	N/A	60	3.97	28.02
SS + Ti/TiN	3	40	1.36	9.60
Baseline SS	N/A	40	4.93	34.80

*Erosion rate refers to the specific test case with a feed rate of 8.5 g/min (0.14 g/sec) of Al_2O_3.
**Erosion rates were too low for accurate measurement using thin coating test methodology.

Multilayer TiCrN/TiCr cermet coatings deposited by the LAFAD process have demonstrated a superior corrosion resistance along with low stresses, exceptionally good adhesion and high toughness. Further improvement was achieved by using a 2-segment TiCrN/Ti+TiBCN coating deposited by hybrid LAFAD+UBM process. The results of salt fog simulated corrosion testing of high chromium Pyrowear 675 steel with 6 μm and 12 μm LAFAD TiCrN/Ti coatings have demonstrated almost complete elimination of pitting corrosion, while two- segment TiCrN/Ti(6μm)+BCN(2μm) hybrid coating have demonstrated complete inhibition of pitting corrosion.

Figure 11. Surface of 70μm hybrid $YSZ+(Al,Cr)_2O_3$ LAFAD+EBPVD coating after (500-1100°C) thermal cycling testing.

Thick oxi-ceramic coatings such as yttria-stabilized zirconia (YSZ) are used as thermal barrier coatings (TBCs) for turbine blades. TBC deposition from near neutral vapor in conventional EBPVD results in large columnar grain morphology which is detrimental both for oxidation and other chemically bounded degradation in reactive atmospheres, and often encourages excessive impact wear. Therefore, deposition of TBC with fine grain structure is considered one of the ways for improving the protective ability of TBC for turbine blades applications. Multilayer, nano-structured oxi-ceramic coatings consisting of the YSZ layers in turn with $YSZ+(Al,Cr)_2O_3$ nanocomposite layers were deposited on Hastelloy-X substrates using the highly ionized vapor plasma generated by hybrid LAFAD+EBPVD process. The results of thermal cycling test have demonstrated exceptional thermal-chemical stability of this coating. The average grain size did not exceed 0.5μm after thermal cycling as shown in Figure 11; this is a marked improvement over the behavior of conventional (non-ionized) EBPVD coatings[21].

CONCLUSION AND CONSIDERATIONS FOR FUTURE WORK

LAFAD and hybrid filtered arc assisted vapor deposition technologies utilizing the LAFAD process in combination with conventional EBPVD and/or UBM processes have demonstrated the ability to deposit dense, nano-structured, cermet and ceramic coatings with low defect density and high thermochemical stability. The resulting coatings have nanostructured, multilayer architectures with ultra-fine grain morphology, excellent adhesion and corrosion resistance. Excellent high temperature stability in oxidizing environments has been demonstrated during thermal cycling to temperatures as high as 1100°C. The coatings have already demonstrated substantial improvement of erosion resistance to impacts in the moderate kinetic energy impact range, and it is expected that significant improvement of erosion resistance to impacts in the high kinetic energy impact range will soon be achieved as the technology for depositing thicker coatings matures. The deposited cermet coating effectively encapsulates the metal substrate, thereby reducing impact-induced surface deformation, and the nanolayered structure effectively inhibits the propagation of cracks in the direction perpendicular to the surface. Future work will be focused on selection of the coating composition and architecture, using correlation between single impact erosion scar pattern and gravimetrical analyses of mass losses using different erosion media, different velocities and feed rates. The coatings will be optimized to provide erosion protection, while retaining corrosion and high temperature oxidation resistance before the substrate is exposed.

ACKNOWLEDGEMENT

The authors would like to acknowledge the technical assistance of Recep Avci and Richard Smith at Montana State University. Thanks to Dave VanVorous for carrying out the coating deposition trials. Thanks are also due to Mr.Chuck Blair for erosion resistant testing at UDRI. Portions of this research were supported by the United States Department of Defense via the program No. W911NF-05-2-0016.

REFERENCES

1. A.W. Ruff and L.K. Ives, "Measurement of Solid Particle Velocity in Erosive Wear," Wear **35** (1975) p.195.
2. Manish Roy, "Elevated temperature erosive wear of metallic materials," J. Phys. D: Appl. Phys. 39 (2006) R101–R124.
3. A.Sue, Surface and Coating Technology, 62 (1993), 115-120.
4. Shanov, W.Tabakoff, Surface and Coating Technology, 86-87 (1996) 88-93.
5. Hamed, W. Tabakoff, R. Wenglarz, "Erosion and Deposition in Turbomachinery," J. Propul. Power, Vol. 22, No. 2, March–April 2006.
6. Jean-Pierre Immarigeon, David Chow, V.R. Parameswaran, Peter Au, H. Saari and Ashok K. Koul, "Erosion Testing of Coatings for Aero Engine Compressor Components," Advanced Performance Materials, Volume 4, Number 4 / October, 1997, p.371-388.
7. V.R Edwards and P.L Rouse. U.S. Army Rotorcraft Turboshaft Engines Sand and Dust Erosion Considerations. Erosion, Corrosion and Foreign Object Damage Effects in Gas Turbines. AGARD CP 558.
8. R Hauret and J Patscheider. From Alloying to Nanocomposites - Improved Performance of Hard Coatings. Advanced Engineering Materials **2**, 247-259.
9. P.Y Yashar and W.D Sproul. Nanometer Scale Multilayer Hard Coatings. Vacuum **55**, 179-190 (1999).
10. Goat, C. Erosion Resistance in Metal-Ceramic Multilayer Coatings for Gas Turbine Compressor Applications. 94. Cranfield , Cranfield Univeristy.
11. Gachon, Y., Vannes, A. B., Farges, G., Sainte Catherine, M. C., Caron, I., and Inglebert, G. Study of sand particle erosion of magnetron sputtered multilayer coatings. Wear 233-235, 263-274. 99.
12. G.Y.Richardson, C.S.Lei, and W.Tabakoff, International Journal of Rotating Machinery, 9(1): 35-40, 2003.
13. Zhao Linruo, "Nanostructured Coatings for Erosion Protection of Engine Components", Seminar on Advanced Coatings and Surface Systems for Application in Aerospace, Marine and Transportation Industries, The Singapore Institute of Manufacturing Technology, 25 February 2003.
14. Vladimir I. Gorokhovsky , Rabi Bhattacharya and Deepak G. Bhat, Surface and Coating Technology, **140** (2) 2001, pp. 82-92.
15. V.I. Gorokhovsky, P.E. Gannon, M.C. Deibert, R.J. Smith, A. Kayani, M. Kopczyk, D. VanVorous, Z. Gary Yang, J.W. Stevenson, S. Visco, C. Jacobson, H. Kurokawa, S.W. Sofie, Journal of The Electrochemical Society, 153 (10) A1886-A1893 (2006).
16. V.Gorokhovsky, C.Bowman, P.Gannon, D.VanVorous, A.A.Voevodin, A.Rutkowski, C.Muratore, R.J.Smith, A.Kayani, D.Gelles, V.Shutthanandan, B.G.Trusov, Tribological performance of hybrid filtered arc-magnetron coatings. Part I, Surface and Coatings Technology, 201 (2006) 3732-3747.
17. I.G. Wright, "Is there any reason to continue research efforts in erosion-corrosion?" *Proc. John Stringer Symposium on High Temperature Corrosion*, ASM International, 2001.
18. In-Wook Park, Kwang Ho Kim, Augusto O.Kunrath, Dalong Zhong, Jong J. Moore, A.A.Voevodin, E.A.Levashov, Microstructure and mechanical properties of superhard Ti–B–C–N films deposited by dc unbalanced magnetron sputtering J.Vac.Sci.Technol. B Vol. 23, No.2, Mar/Apr 2005 p.588-593.
19. H. Jehn, G. Reiners, N. Siegel, DIN-Fachbericht (Special Report) 39, Charakterisierung duenner Schichten (Characterization of thin layers), Beuth-Verlag, Berlin, 1993.
20. N. Novikov, V.I. Gorokhovsky and B. Uryukov, Superhard i-C Coatings Used in Complex Processes of Surface Strengthening of Tools and Machine Parts, Surf. Coat. Techn., **47**, 770(1991).
21. B.-K.Jang, H.Matsubara, Materials Letters vol. 59 (2005) 3462-3466.

Coatings to Resist Wear and Tribological Loadings

DEPOSITION AND CHARACTERIZATION OF DIAMOND PROTECTIVE COATINGS ON WC-CO CUTTING TOOLS

Y. Tang, S.L. Yang, W.W. Yi, Q. Yang
Department of Mechanical Engineering, University of Saskatchewan,
57 Campus Drive, Saskatoon SK S7N 5A9, Canada

Y.S. Li, A. Hirose
Plasma Physics Laboratory, University of Saskatchewan,
116 Science Place, Saskatoon, SK S7N 5E2, Canada

R. Wei
Southwest Research Institute
6220 Culebra Road, San Antonio, TX 78228, U.S.A.

ABSTRACT

Diamond coatings were deposited on WC-Co cutting inserts by microwave plasma enhanced chemical vapour deposition. To restrict the catalytic effect of cobalt binder phase for graphite formation, a combination of hydrochloric acid etching and H_2 plasma treatment was used as substrate pretreatment to reduce surface Co concentration and modify substrate surface morphology. SEM results show that continuous nanocrystalline diamond coatings were formed on pretreated substrates, whereas a composite of diamond and graphite was formed on untreated substrates. Raman spectra show strong diamond characteristic peak at about 1332cm^{-1} on pretreated substrates, while only strong graphite peaks and weak diamond peak are on untreated substrates. No film spallation was observed along the scratching track in scratching tests of the diamond coatings on pretreated substrates. Among the three tested pretreatments (HCl+H_2, H_2+HCl, HCl), H_2+HCl, hydrogen plasma etching followed with hydrochloride etching produces the highest diamond phase content.

INTRODUCTION

With sp^3 hybridization and strong symmetric covalent bonding, diamond is by far the hardest known natural material. Diamond has a low coefficient of thermal expansion and a high thermal conductivity. It is chemically inert, has low friction coefficient, and excellent wear resistance. The unique properties of diamond as well as the recent development of chemical vapour deposition (CVD) diamond coating make diamond an ideal protective coating for cutting tools. The most suitable cutting tool material for the growth of diamond coating is cobalt-cemented tungsten carbide (WC-Co). WC-Co has been widely used in the machining of carbon/graphite, plastics, fibre glass, and aluminium composite. Applying diamond coatings on the tool surfaces is expected to prolong the WC-Co cutting lifetime and improve the manufacturing efficiency. However, CVD diamond coatings suffer from severe premature adhesion failure on cements tools due to the interfacial graphitization induced by the binder phase Co. Co-cemented carbides contain cobalt as a binder, which provides additional toughness to the tools, but it is hostile to diamond adhesion[1-3]. Co, being a transition metal with partially filled *3d* shell, acts as a catalyst for the formation of graphite[4]. The graphite layer formed during the early stage of deposition makes diamond difficult to well adhere to the underlying substrate[5].

There are two widely used approaches to reduce the deleterious effect of cobalt and improve the adhesion of diamond coating on WC-Co. The first one is to suppress the interactions between the binder and the deposited carbon films by using an interlayer which has a low diffusion coefficient for both C and Co and an intermediate thermal expansion coefficient between WC-Co and diamond. Many researchers have reported improvements in adhesion using various interlayers, such as amorphous carbon[6], metallic materials (Ti-Si)[7], ceramics (SiAlON, Al_2O_3, SiC, Si_3N_4, TiC, TiN and WC)[8-11] and whisker reinforced ceramics (Al_2O_3 + SiC and ZrO_2)[8]. The other way to enhanced diamond coating adhesion is to remove the cobalt binder phase on the surface by chemical etching. Peters and Cummings[12] developed an effective two-step chemical etching process in which the substrate was first treated with Murakami's reagent ($K_3Fe(CN)_6$: KOH : H_2O = 1 : 1 : 10) and then etched with Caro's acid (H_2SO_4–H_2O_2 solution). Murakami's reagent attacks WC grains and thus roughens the substrate surface whereas Caro's acid oxidizes the binder to soluble Co^{2+} compounds and thus reduces the surface Co concentration. Chemical etching is a more cost-saving method than applying interlayers. However, $(CN)^{-1}$ in the reagent is extremely hazardous for human beings and the environment. It is necessary to develop a cleaner pretreatment method for the growth of adhesive diamond coatings on WC-Co. In this work, a combination of hydrochloric acid etching and H_2 plasma treatment were used as pre-treatment to reduce surface Co concentration and modify the substrate surface morphology to increase coating adhesion.

EXPERIMENTAL

In order to decrease the Co content on the WC-Co surface, two methods, hydrochloric (HCl) acid etching and hydrogen (H_2) plasma treatment, were used in this experiment to etch Co out and modify the surface of WC-Co. For HCl etching, the WC-Co substrates were immersed in HCl solution (30% v/v) for 40 minutes to form an "etched layer" with a thickness of a few micrometers and then rinsed with deionised water. The H_2 plasma treatment was using a 2.45 GHz Microwave plasma CVD reactor manufactured by Plasmionique Inc as depicted somewhere else[13]. The vacuum chamber was pumped down to a pressure of 6.65×10^{-4} Pa using a turbo-molecular pump and then filled with hydrogen (H_2). The flow rate of H_2 was 100sccm. When the working pressure was stabilized at 4 kPa, a 2.45 GHz microwave source was switched on to form a plasma. The working gas pressure was maintained at 4 kPa and the microwave power 1000 W. During the treatment, the substrates were heated only by the plasma and the substrate temperatures were about 570°C as measured with a thermocouple mounted right behind the substrate holder. The duration of treatment was 4 hours.

Three different combinations of chemical etching and plasma treatment were applied as pretreatments: (1) HCl etching, (2) H_2 plasma treatment +HCl etching, and (3) HCl etching + H_2 plasma treatment.

The synthesized materials were characterized using scanning electron microscopy (SEM) and Raman spectroscopy. The Raman spectra were obtained using a Renishaw micro-Raman system 2000 spectrometers operated at a laser wavelength of 514.5 nm generated by an argon laser. The spot size was approximately 2 μm. The adhesion of diamond coatings on WC-Co was examined by scratch testing.

RESULTS AND DISCUSSION

The surface morphologies of WC-Co substrates before and after pretreatment are shown in Fig. 1. Fig. 1a shows the original WC-Co surface. Grooves can be found on the surface because the cutting inserts were finished with grounding by the manufacturer. The bright phase is WC and the dark phase is Co. After HCl solution etching, as shown in Fig. 1b, Co phase on the surface was partly etched out and the surface became roughened. Fig. 1c shows the surface morphology of WC-Co pretreated with H_2 plasma followed by HCl etching. The surface treated by this combination seems more effective to remove Co and has a rougher surface than treated only by HCl etching. In the third pretreatment method, the sequence of the combination was changed, that is, the substrates were first treated with HCl and then with H_2 plasma. As shown in Fig. 1d, it seems that this pretreatment is less effective to remove Co and small Co particles can be seen on the surface of Co and WC grains. This may attribute to the diffusion of Co onto the surface due to the heat in the H_2 plasma treatment.

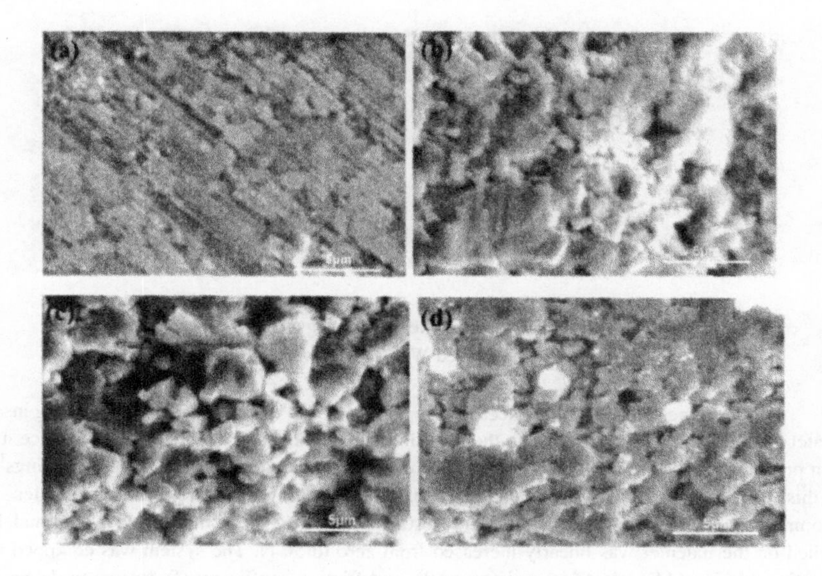

Fig.1 SEM surface morphologies of (a) non-pretreated and pretreated WC-Co cutting inserts with (b) HCl etching (c) H_2 plasma+ HCl etching and (d) HCl etching+ H_2 plasma.

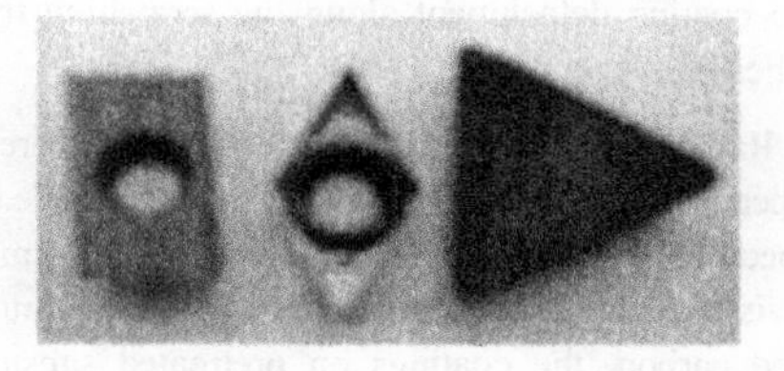

Fig.2 Optical image of different WC-Co cutting inserts coated with diamond.

All the as-deposited coatings on pretreated and non-pretreated cutting inserts were first examined by optical microscopy. It was found that all the coatings on pretreated substrates were smooth, uniform and continuous. Fig. 2 shows the typical optical image of different WC-Co cutting inserts coated with diamond. It was also found that the coatings on non-pretreated substrate were coarse and non-continuous.

The surface morphologies of coatings were further characterized by SEM, as shown in Fig.3. Fig.3a shows the materials grown on non-pretreated substrates. It can be seen that a composite of diamond and graphitic crystals was formed on the surface. The binder phase Co is responsible for the formation of graphite. Fig.3b shows the typical surface morphology of the coating grown on pretreated substrates. One can find that continuous nanocrystalline diamond coatings were formed on pretreated substrates.

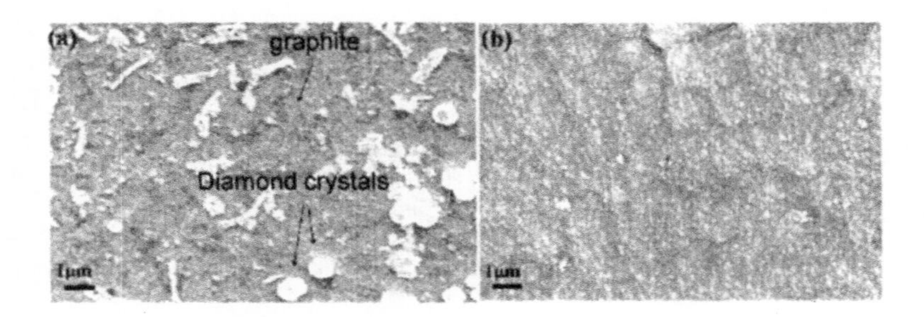

Fig.3 Typical SEM surface morphologies of coatings (6 hrs) on (a) untreated and (b) pretreated WC-Co substrates.

Coating adhesion is a main concern in the application of diamond on WC-Co cutting inserts. Scratch testing was chosen to evaluate the adhesion between the coating and substrate, since it has been proved to be the most practical method for evaluating the adhesion of hard, thin coatings[14-15]. In this experiment, scratch testing of the diamond coatings on pretreated substrates was accomplished with a Universal Micro/Nano testing system. In the scratch test, the normal load applied on the indenter was linearly increased from zero to 50 N. The system was equipped with acoustic emission (AE), electrical resistance (R), and friction coefficient (f) sensors to detect film cracking and spalling. There was no evidence of coating detachment up to 50 N normal load from the AE, R and f diagram. The surface morphology of scratched diamond coatings was examined with SEM as shown in Fig.4: 4a with lower magnification and 4b with higher magnification. It can be seen that there was no coating detachment along the scratching track during scratch testing, which indicates a good adhesion.

Fig. 5 shows the Raman spectra of diamond coatings on pretreated and non-pretreated substrates. Characteristic peak of diamond at about 1332cm^{-1} can be clearly defined in all pretreated WC-Co substrates. The spectra also show a broad peak around 1580 cm^{-1}. Considering the fact that Raman scattering in the visible range (514 nm) is about 70 times more sensitive to sp^2-bonded carbon than to sp^3-bonded carbon, the coatings on pretreated substrates have high content of diamond. Comparing the spectra of samples with different pretreatments, one can find that the diamond coating on H_2+HCl pretreated substrate has the highest and sharpest diamond characteristic peak, which indicates highest diamond content. HCl and HCl+H_2 pretreated samples have comparable diamond content. The diamond peak of non-pretreated samples is rather weak and graphite peaks are strong. It indicates that the coating has very high content of non-diamond carbon.

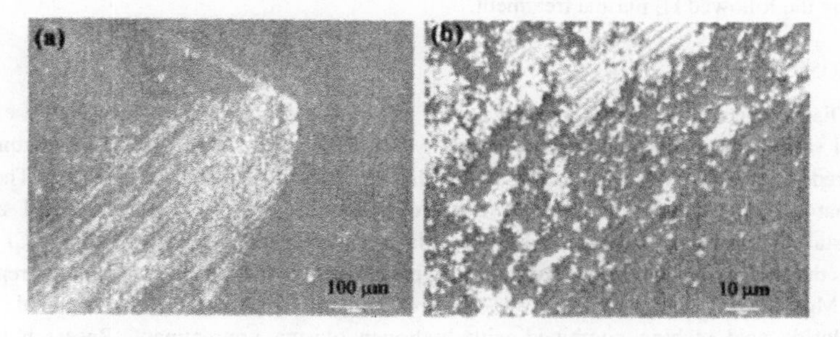

Fig.4 SEM surface morphologies of diamond coatings on pretreated substrate after scratching test.

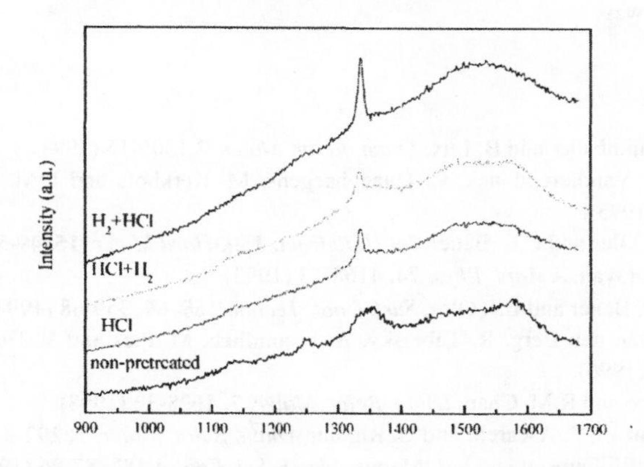

Fig. 5 Raman spectra of diamond coatings on pretreated and untreated substrates.

From the results, it can be obviously seen that the three pretreatments, HCl, $HCl+H_2$, and H_2+HCl, are all effective in improving the diamond phase content and coating adhesion. Hydrochloride solution as a strong acid reacts and etches out the Co binder phase on the substrate surface. Then the forming of graphitic phase induced by Co will be restricted. Hydrogen plasma pretreatment has two effects on WC-Co substrates according to the research by Li et al[16]. One is that the heating of H_2 plasma results in the growth of WC crystal grains. However, coarsening of WC grain was not clearly shown from the SEM results in this study. The other effect of H_2 plasma pretreatment is that H_2 plasma can decarbonized substrate surface and reduce the chemical state of tungsten. The reduced tungsten carbides then react with carbon atoms from precursor gases during the following deposition. The reaction can not only result in chemical bonding between diamond coatings and substrates but also improve the diamond nucleation on WC-Co. From Fig.1, we can see that hydrogen plasma treatment followed by acid etching is more effective to remove the Co in the sample surface and produces rougher surfaces, which increases the nucleation density of diamond and the actual contact area of coating/substrate, thus improves the coating quality and the adhesion of diamond coatings. However, it should be noticed that $HCl+H_2$ pretreatment is not as effective as H_2+HCl pretreatment. The reason can be attributed to that Co atoms underneath the surface diffused out during the followed H_2 plasma treatment.

CONCLUSION

Diamond coatings were deposited on the WC-Co inserts by microwave plasma enhanced chemical vapour deposition. A combination of hydrochloric acid and H_2 plasma pretreatment was used to reduce surface Co concentration and modify the substrate surface morphology. The results show that the tested pretreatments were all effective in forming continuous and adhesive nanocrystalline diamond coatings, while H_2 plasma pretreatment followed with HCl etching produces the highest diamond phase content. The preliminary results hold the promise to replace the hazard Murakami's regent in the WC-Co pretreatment with more environmental friendly hydrochloride acid etching combined with hydrogen plasma pretreatment. Research work on precisely measure the critical load of diamond coatings by scratch testing and evaluate the diamond phase by near edge X-ray absorption fine structure as well as cutting performance of coated inserts are currently underway.

REFERENCES

[1] R. Haubner, A. Lindlbauer and B. Lux, *Diam. Relat. Mater.* **2** 1505-15 (1993).
[2] M. Nesládek, K. Vandierendonck, C. Quaeyhaegens, M. Kerkhofs and L.M. Stals, *Thin Solid Films* **270,** 184-8 (1995).
[3] A. Inspektor, E.J. Oles and C.E. Bauer, *Int. J. Refract. Met. Hard Mater.* **15,** 49-56 (1997).
[4] X. Chen and J. Narayan, *J. Appl. Phys.* **74,** 4168-73 (1993)
[5] A. Inspektor, C.E. Bauer and E.J. Oles, *Surf. Coat. Technol.* **68–69,** 359-68 (1994).
[6] F. Deuerler, H. van den Berg, R. Tabersky, A. Freundlieb, M. Pies and V. Buck, *Diam. Relat. Mater.* **5,** 1478-89 (1996).
[7] C.R. Lin, C.T. Kuo and R.M. Chan, *Diam. Relat. Mater.* **7,** 1628-32 (1998).
[8] E. Cappelli, F. Pinzari, P. Ascarelli and G. Righini, *Diam. Relat. Mater.* **5,** 292-8 (1996)
[9] J.K. Wright, R.L. Williamson and K.J. Maggs, *Mater. Sci. Eng. A* **187,** 87-96 (1994).
[10] S. Silva, V.P. Mammana, M.C. Salvadori, O.R. Monteiro and I.G. Brown, *Diam. Relat. Mater.* **8,** 1913-8 (1999).
[11] C. Faure, W. Hänni, C. Julia Schmutz and M. Gervanoni, *Diam. Relat. Mater.* **8,** 830-3 (1999).
[12] M.G. Peters and R.H. Cummings, European Patent 0519587 A1 (1992)
[13] W. Chen, X. Lu, Q. Yang, C. Xiao, R. Sammynaiken, J. Maley and A. Hirose, Thin Solid Films **515,** 1970-5 (2006)
[14] P.A. Steinmann and H.E. Hintermann, J. Vac. Sci. Technol. **A7,** 2267–72 (1989)
[15] H. Ollendorf and D. Schneider, Surf. Coat. Technol. **113,** 86–102 (1999)
[16] J.G. Li, D.P. Hu, J. Mei. S. Liu and Y.Y. Li, Acta Metallur. Sinica **42,** 763-9 (2006)

FRICTION AND WEAR BEHAVIOR OF ZIRCONIA CERAMIC MATERIALS

C. Lorenzo-Martin, O. O. Ajayi, D. Singh, and J. L. Routbort
Energy Systems Division
Argonne National Laboratory
Argonne, IL 60439

ABSTRACT

Zirconia is one of the most versatile structural ceramic materials because of the range of possible microstructures, and hence mechanical properties. Toughened zirconia materials with different microstructures are currently used for tribological applications either in the monolithic forms or as coatings. In the present study, lubricated friction and wear behaviors of five different zirconia materials with different microstructures were evaluated using block-on-ring contact configuration and unformulated PAO lubricant. Tests were conducted with zirconia blocks sliding against hardened steel coated with hard MoN thin film. Although the friction behaviors of the all the tested materials are similar with average friction coefficient of 0.10 – 0.12 for all materials, significant differences were observed in their wear attributes. About a factor of 10 difference was observed in the wear rates of the materials. Although the behavior cannot be easily connected to any particular material property, in general, materials with multi-phase microstructure exhibited less wear than mono-phased materials.

INTRODUCTION

Structural ceramic materials are increasingly used for tribological applications. For instance, silicon nitride (Si_3N_4) is used as rolling elements in hybrid bearing, zirconia (ZrO_2) is used as fuel injector plunger in heavy duty diesel engines; and silicon carbide (SiC) is used as mechanical face seals. These materials possess combination of properties, such as high hardness, that are desirable for tribological application. Although the friction and wear performance of structural ceramic materials have been studied for some time, there is still no clear guideline for selection of these materials for a particular tribological application, except through empirical testing.

In some of the earlier studies of ceramic wear, attempts were made to connect wear performance with some of the mechanical properties of the material, especially hardness and fracture toughness, in an attempt to devise a selection guideline [e.g., 1,2]. Other studies showed that the wear of ceramic material is a very complex process and simple connection with large scale conventional mechanical properties, especially fracture toughness and hardness may not be possible [e.g. 3]. Nonetheless, it is desirable to optimize the wear resistance of ceramic materials and if possible to predict the wear behavior for particular contact conditions or applications. For more extensive use of ceramics for tribological applications, there is the need for selection and design guideline for ceramic tribo components.

The tribological behavior of steel is well studied because the vast majority of tribological components are made of ferrous materials in the form of steel or cast iron. It is well known that the friction and wear of steel is to a very large extent dependent on their microstructure. Indeed a common

* Work supported by the Department of Energy, under Contract DE-AC02-06CH11357

way to modify the wear behavior of steel is through microstructural modification, usually accomplished by heat treatment. Some studies have shown that microstructure has an impact on the wear of alumina (Al_2O_3) [e.g., 4,5], indicating that the impact of microstructure on wear is not limited to steel. Among structural ceramics, zirconia (ZrO_2) is perhaps the most versatile in terms of microstructure. With the addition of variety of stabilizing dopants and heat treatments, ZrO_2 can be fabricated with a variety of microstructures in a manner analogous to steel.

The goal of the present study is to evaluate the friction and wear behavior of structural ZrO_2 materials with different microstructures. The results of the study may provide insight into the microstructural features of the material pertinent to their tribological behavior.

EXPERIMENTAL DETAILS:

ZrO_2 Materials:

The friction and wear behaviors of five different commercially available ZrO_2 were evaluated in this study. The materials tested and some of their properties and microstructural characteristics are shown in Table I. The phase constituent of each material shown in the table was determined by x-ray diffraction analysis. Microstructures and grain morphology for each material were evaluated by SEM of the polished and etched (HF) sample. The Y-TZP material (Z) consists primarily of tetragonal phase and some monoclinic phase most likely formed by phase transformation on the sample surface during preparation by grinding (Fig. 1). The SEM micrographs of etched samples for the five materials are shown in Figure 2. Y-TZP consisted of uniaxial, homogeneously distributed tetragonal grains of about 200 nm grain size (Fig. 2a). The MS material consists of relatively large cubic grains (30 μm) with uniform fine tetragonal precipitates (Fig. 2b). The TS material also consisted of cubic phase (30 μm) with both primary (150 nm) and secondary (30 nm) tetragonal precipitates (Fig. 2c). Similarly, the L material also consists of cubic grains with tetragonal precipitates. The primary precipitate was only about 50 nm while the secondary tetragonal precipitate was about 30 nm (Fig. 2d). The Y material is a fully established cubic material with a grain size of about 25 μm, and small amount of about 2-3 μm monoclinic particles as shown in Figure 2e.

Friction and Wear Testing:

Friction and wear tests were conducted using block-on-ring contact configuration as shown schematically in Fig. 3. The stationary 15 x 10 x 6.3 mm block specimens are made of the ZrO_2 material, while the rotating 34 mm outer diameter ring is made of 4620 steel with a 2 μm hard MoN thin film coating. The coating has a hardness of about 30 GPa compared to the hardness of 11-12 GPa for most of the ZrO_2 test materials. The significant differences in the hardness of block and ring materials will ensure that bulk of wear will occur on the ZrO_2 blocks. All the tests were lubricated with synthetic basestock poly alpha olefin (PAO-4) oil with viscosity of 18 cSt at 40°C. Unformulated lubricant was used so as to eliminated the chemical impact of lubricant on tribological performance.

Wear tests were conducted with a normal load of 50 N which imposed a Hertzian line contact of about 120 MPa; ring rotating speed of 750 rpm which produces a sliding velocity of 1.3 m/s at the sliding contact interface. Each test was conducted for duration of 30 minutes during which the normal and tangential loads were continuously monitored by a three axes load-cell. The friction coefficient, which is the ratio of the tangential to the normal forces, was hence continuously monitored. For each material, a minimum of three tests were conducted to assess the repeatability of the tribological behavior of the materials. At the conclusion of the test, the amount of wear in the ZrO_2 block material was measured by optical profilometry technique as shown in Figure 4. The volume of material removed by wear was directly calculated by the profilometer software. The wear mechanisms in each material were also assessed by post-test SEM examination.

RESULTS AND DISCUSSION:

The typical variation of friction coefficient with time during testing for each material is shown in Figure 5. The frictional behavior is very repeatable for each material. Similarly, the time variations of friction coefficient for the five materials are quite similar to one another. The only subtle difference in the frictional behavior of the materials is the level of frictional noise. The friction coefficient was quiet noisy in all the tests with the TZP material (Figure 5a) and much less so in the tests with the PSZ (L) material (Figure 5b). This frictional behavior is perhaps due to differences in wear mechanisms of the materials as will be discussed later. The differences in frictional behavior could have a significant technological implication for material selection for tribological applications. Materials with noisy frictional behavior will not be appropriate for high precision sliding and components that must operate quietly.

The magnitudes of friction coefficient for all the material were very similar to one another and ranges from 0.10 to 0.12 as shown in Table II. This friction coefficient magnitude is typical for boundary lubrication regime in which the lubricant fluid film thickness is not large enough to completely separate contacting sliding surfaces. Under such condition, friction is due to the shearing of the fluid film and contact points between the two sliding surfaces.

Although the frictional behaviors of the five materials were similar to one another in both trend and magnitude, there were significant differences in their wear behavior. Table III and Figure 6 show the measure wear volume in each material at the conclusion of test. Since all the tests were conducted under the same contact conditions of load, speed and duration, the total sliding distance is the same, hence the relative wear volume between the materials also reflects their relative wear rate (or wear resistance). There is an order of magnitude difference the material with lowest wear (L) and the one with highest wear (Y-TZP). Under our test conditions, there is clearly no discernable connection between the wear rate and mechanical properties of the materials. For instance, in the three partially stabilized materials (MS, TS, L), both their hardness and fracture toughness are similar, but there are significant differences in their wear rates. Also, the Y material has the lowest hardness and fracture toughness, but a better wear resistance than Z and TS.

The differences in the wear rate can be attributed to differences in wear mechanisms occurring in each of the materials. For the TZP material, the predominant wear mode is polishing as shown in Figure 7. Material removal occurs at individual grain scale, which is relatively small. Consequently some of the loosened grain are entrained into the contact interface with the lubricant. The presence of such wear particles within the contact interface is expected to produce noisy frictional behavior as shown in Figure 5a. Wear mechanisms in the three PSZ materials also involved polishing, but in addition, localized spall damage as illustrated in Figure 8b. Because two phases are involved, the rate of material removal by polishing mechanism will not be the same for both phases, which can result in the reduction of wear rate compared to a monophase material. Also for the PSZ materials, transformation of some of the tetragonal precipitate to monoclinic phase under the imposed contact stresses is expected. This is expected to reduce the wear rate of this group of materials because phase transformation is an energy dissipation mechanism and hence will reduce energy available for wear damage. One can attribute differences in the wear rates between the three PSZ materials to differences in their tetragonal phase precipitate morphology, distribution and perhaps transformability Wear in the cubic material occurs primarily by polishing of the large grains and pluck out of much smaller monoclinic grains, usually in the grain boundary region as shown in Figure 9.

The lack of connection between the conventional mechanical properties and wear behavior of the ZrO_2 was observed under the test conditions in the present study. The mechanical properties measure macro scale events, for instance hardness measures a material's resistance to large scale plastic deformation and fracture toughness measures the material's resistance to large and fast crack propagation. It is thus not surprising that the wear in our study which occurs at a micro level or scale at the sliding contact interface is more connected to the material microstructure rather than macro properties. Under more severe contact conditions, in which a change in wear mechanism is possible [e.g., 6,7], it may be possible to empirically relate wear behavior and some mechanical properties.

SUMMARY AND CONCLUSIONS:

The friction and wear behavior of five commercially available structural ZrO_2 ceramic materials were evaluated in a PAO basestock lubricated contact with MoN coated steel using block-on-ring contact configuration. The friction coefficient trend and magnitude for all the five materials are similar to one another. The measured friction coefficient of 0.10 to 0.12 range is typical for boundary lubrication regime. In spite of similarity of frictional behavior the wear rates in the materials are significantly different. An order of magnitude difference in the wear rate was observed. The differences in wear rate reflect differences in the operating wear mechanisms in materials with different microstructures. No obvious empirical connection can be established between the wear rate and mechanical properties notably fracture toughness and hardness.

ACKNOWLEDGEMENT

This work was supported by the Office of Vehicle Technologies of the U.S. Department of Energy under contract DE-AC02-06CH11357.

REFERENCES:

[1]T. E. Fischer, M. P. Anderson and S. Jahanmir, "Influence of Fracture Toughness on Wear Resistance of Titria Doped Zirconium Oxide," J. Am. Ceram. Soc. 72 252-257 (1989).
[2]A. G. Evans and D. B. Marshall, "Wear Mechanisms in Ceramics," pp. 439-52 in Fundamentals of Friction and Wear of Materials, Ed. By D. A. Rigney, ASM (1980).
[3]C. P. Dogan and J. A. Hawk, "Microstructure and Abrasive Wear in Silicon Nitride Ceramics," Wear 250, 256-263 (2001).
[4]K.-H. Zum Gahr, W. Bundschuh and B. Zimmerlin, "Effect of Grain Size on Friction and Sliding Wear of Oxide Ceramics," Wear, 162-164, 269-279 (1993).
[5]O. O. Ajayi and K. C. Ludema, "The Effect of Microstructure on Wear Modes of Ceramic Materials," Wear 154, 371-385 (1992).
[6]S. Jahanmir and X. Dong, "Mechanism of Mild to Severe Wear Transition in α-Alumina," J. Tribol. 114, 403-411(1992).
[7]K.-H. Zum Gahr, "Modeling and Microstructural Modification of Alumina Ceramic for Improved Tribological Properties," Wear 200, 215-224 (1996).

TableI: Summary of ZrO_2 materials evaluated.

SAMPLES	Z	MS	TS	L	Y
CHEMISTRY (wt.%)	2.6%Y2O3-TZP	3.4%MgO-PSZ	3.4%MgO-PSZ	3%MgO-PSZ	8% Y2O3-FSZ
MICROSTRUCTURE	tetragonal	Cubic + tetragonal precipitates	Cubic + 1st & 2nd tetragonal precipitates	Cubic + tetragonal precipitates	Cubic
DENSITY (g/cm3)	6.0	5.7	5.7	5.6	5.7
FRACTURE TOUGHNESS (MPa.m1/2)	6.7	8-12	10-15	10	3.5
HARDNESS VICKERS (HV0.3Kg/mm2)	1230	1120	1020	1100	670
SURFACE ROUGHNESS Sa (nm)	335	260	240	230	320

Table II: Average friction coefficient for different materials.

MATERIAL	μ-START	μ-STEADY STATE	μ-AVERAGE
Z	0.13	0.11	0.120 ± 0.010
MS	0.12	0.11	0.115 ± 0.005
TS	0.11	0.10	0.100 ± 0.005
L	0.11	0.115	0.115 ± 0.005
Y	0.13	0.12	0.120 ± 0.005

Table III: Measured wear volume in each ZrO_2 materials

MATERIAL	WEAR VOLUME (10^6 μm^3)
Z	11.4 ± 1.1
MS	4.6 ± 0.8
TS	9.4 ± 0.8
L	1.3 ± 0.1
Y	4.2 ± 0.4

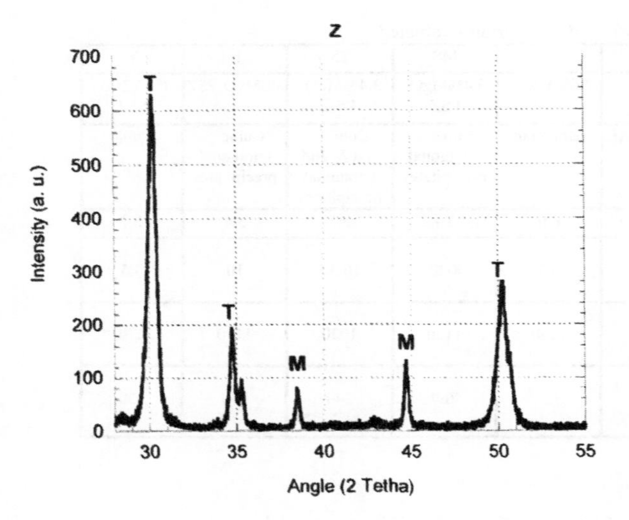

Figure 1: X-ray diffraction spectrum for TZP material showing tetragonal and some monoclinic phase.

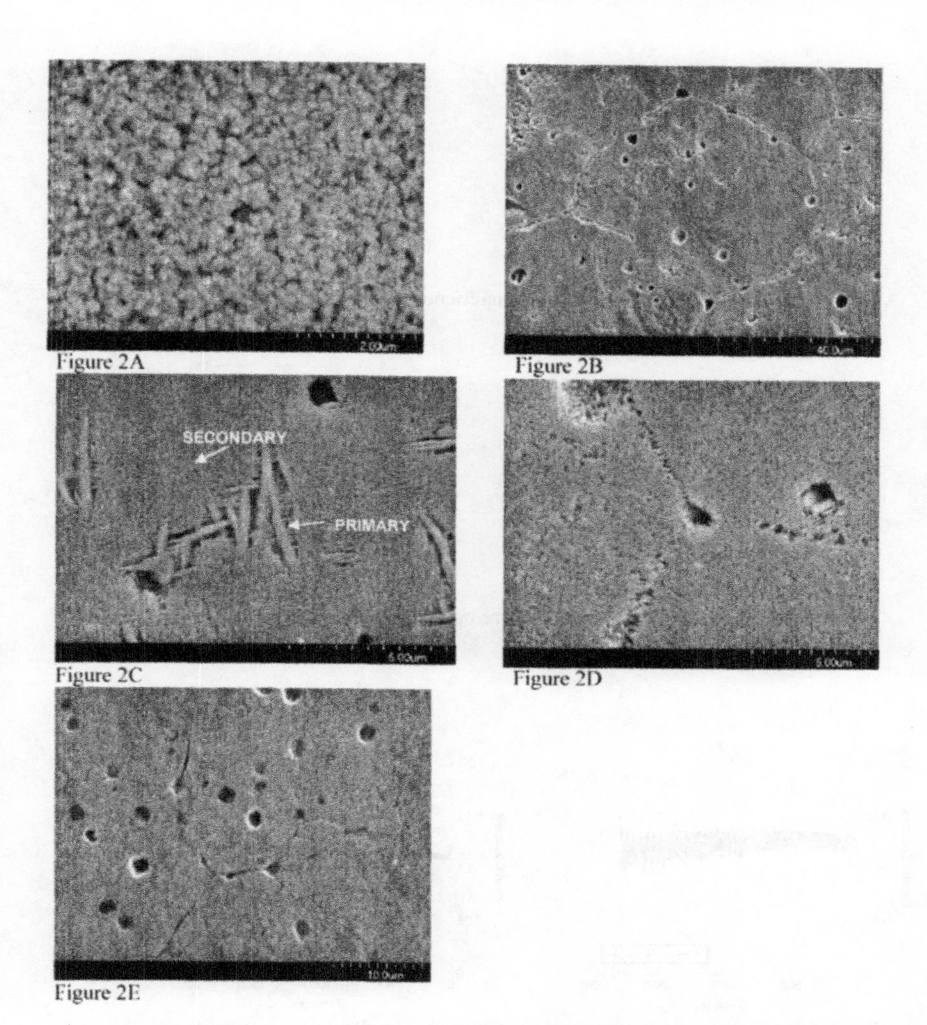

Figure 2: SEM micrograph of HF etched polished samples: a) TZP (Z), b) PSZ (MS), c) PSZ (TS) showing primary and secondary tetragonal precipitates, d) PSZ (L), e) FSZ (Y) cubic.

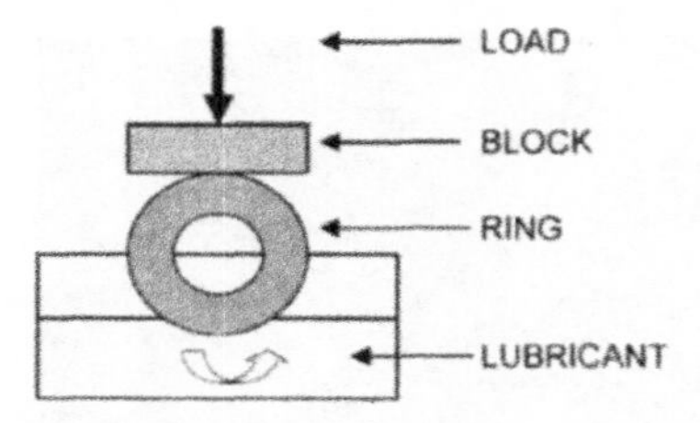

Figure 3: Schematic diagram of block-on-ring friction and wear test system.

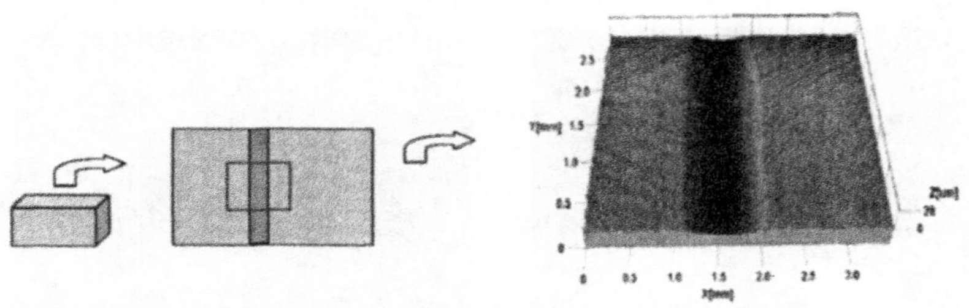

Figure 4: Wear volume measurement with the optical profilometry system.

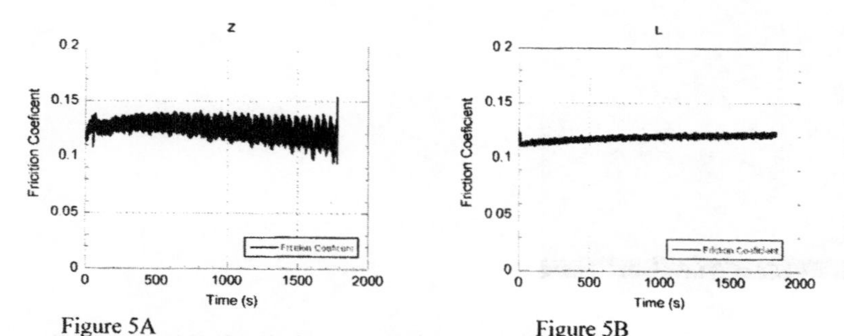

Figure 5A Figure 5B

Figure 5: Variation of friction coefficient with time for: a) TZP (Z) material, b) PSZ (L) material.

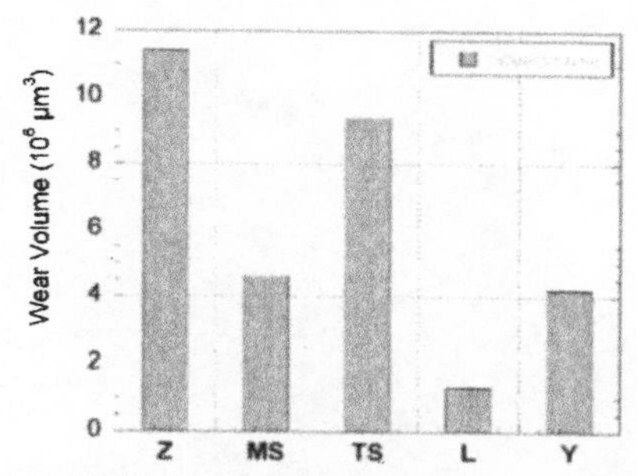

Figure 6: Measured wear volume for the tested materials.

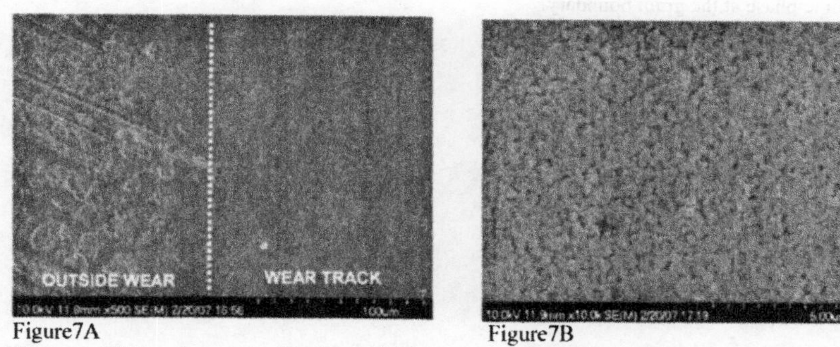

Figure7A Figure7B

Figure 7: SEM micrograph showing polishing wear in the TZP material: a) overall, b) in the wear track.

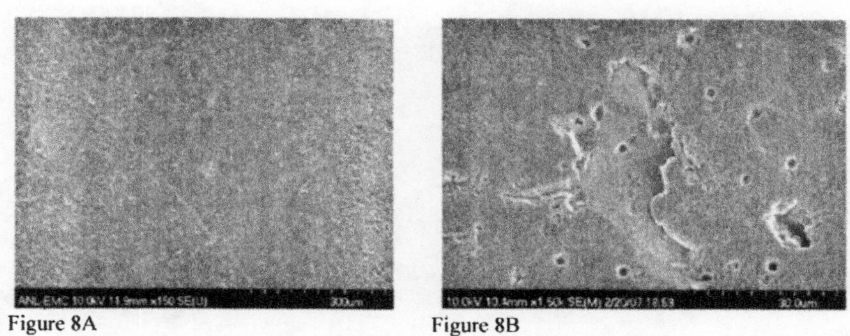

Figure 8A Figure 8B

Figure 8: SEM micrograph of wear track in PSZ material. A) overall wear track and b) localized damage by spalling.

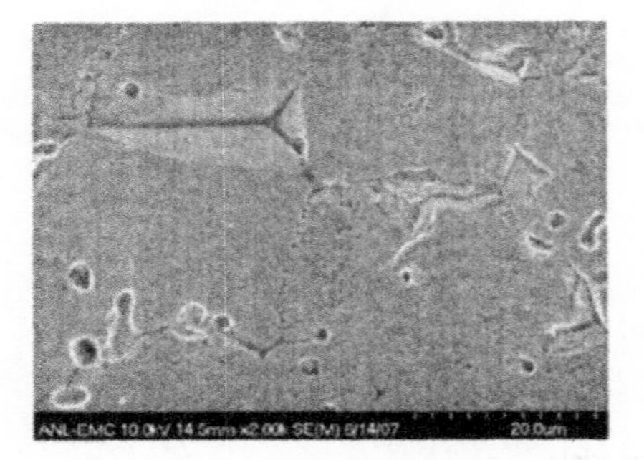

Figure 9: SEM micrograph of wear track in the cubic material (Y) showing polishing and pluck out of monoclinic phase at the grain boundary.

Nanostructured Coatings

CERIUM OXIDE THIN FILMS VIA ION ASSISTED ELECTRON BEAM DEPOSITION

V. Dansoh[1]; F. Gertz[1]; J. Gump[1]; A. Johnson[1]; J. I. Jung[1]; M. Klingensmith[1]; Y. Liu[1]; Y. D. Liu[1]; J. T. Oxaal[1]; C. J. Wang[1]; G. Wynick[1]; D. Edwards[1]; J. H. Fan[1]; X. W. Wang[1]; P. J. Bush[2]; A. Fuchser[3]
1. School of Engineering, Alfred University, Alfred, NY, USA
2. School of Dental Medicine, SUNY, Buffalo, NY, USA
3. J. A. Woollam Co., Inc., Lincoln, NE, USA

ABSTRACT
We report new results on as-deposited cerium oxide thin films with substrate temperatures near 423 Kelvin, via a modified oxygen ion assisted electron beam deposition. In contrast to a conventional ion assisted electron beam deposition process in which the ion beam "milling" and vapor condensation take place simultaneously on a substrate surface, in this process, the ion beam "milling" takes place in a periodically pulsed format while the vapor condensation takes place continuously on a substrate surface. In contrast to an argon ion assisted electron beam deposition process, the oxygen ion source utilized in this process is to fully oxidize the deposited film in situ. The materials properties of the films are analyzed by x-ray diffraction, scanning electron microscopy, energy dispersive X-Ray analysis, Raman image spectroscopy, and variable angle spectroscopic ellipsometry. The film's electrical conductivity is analyzed by through-film impedance spectroscopy from 373K to 1,073K, which enables the determination of the activation energy. A film fabricated by this process appears to have a "sheet-stacking" structure. The coated metallic substrate's tensile stress is measured to make sure that the coating does not negatively affect the mechanical strength of the substrate. The through-film activation energy is measured to be 0.62 ± 0.10 eV, which is comparable to that of the bulk material and some of the thin film materials.

INTRODUCTION
Cerium oxide materials have been studied extensively[1]. Coatings of cerium oxide have been considered in different applications including solid oxide fuel cells[2-3], gas sensors[4], and corrosion protection[5-6]. Thin film cerium oxide materials have been coated on various substrates such as metals, silicon, glass, and ceramics; via different fabrication techniques including spin-coating[7], sputtering[8], pulsed-laser ablation[9-10], electron beam evaporation[11], and ion assisted electron beam depositon[12-16, 34]. Gases utilized in ion sources include oxygen, argon, and mixtures of the two. In a typical ion assisted deposition process, while vapor is condensed onto a substrate, the ion beam is bombarding (milling) the surface of the film simultaneously[15]. In this paper, we report new results with an oxygen ion assisted electron beam deposition process, which is slightly different from a conventional ion assisted electron beam deposition process. The ion beam in this process only "mills" the film in a periodically pulsed format. The additional purpose of utilizing the oxygen ion source is to fully oxidize the deposited film in situ. Traditionally, ion assisted deposition processes are mainly utilized to coat the dense optical films on glass substrates.[15, 23] Instead of coating on glass substrates, our long term goal is to coat an as-deposited dense-protective film on a metallic substrate at a relatively low substrate temperature, without post deposition heat treatment. Thus, the mechanical properties of the coated metallic substrates will not be negatively impacted by the coating or deposition process. Such desire for low substrate temperature coating on the metallic substrate is well documented in the literature[17]. Notice that both CeO_2 and silicon have cubic structures, with similar lattice constants (5.4064 angstrom for silicon, ICDD 01-078-2500, and 5.4150 angstrom for CeO_2, ICDD 00-002-1306). When the alignment of the silicon substrate is <100> or <111>, the alignment of the CeO_2 film may be near <100> or <111>. If the vapor condensation rate is relatively slow, the film growth process may be similar to a "sheet-stacking" process. If a silicon substrate with <111> alignment is chosen, pyramids may be formed. Thus, a silicon substrate with <111> alignment will be chosen to study the film

growth mechanism. Materials properties of the films are analyzed by x-ray diffraction (XRD), scanning electron microscopy (SEM), energy dispersive spectroscopy (EDS), Raman image spectroscopy, and variable angle spectroscopic ellipsometry (VASE). The film's electrical conductivity is analyzed by Impedance Spectroscopy. The coated metallic substrate's tensile stress is measured to make sure that the coating does not negatively affect the mechanical strength of the substrate.

THIN FILM FABRICATION

In Figure 1, an illustration for the ion assisted electron beam deposition system is provided[18]. There are three assemblies: an electron beam assembly located near the lower-center position (Telemark TT-36 power supply/controller, 6 KW, 4 pocket crucible), an oxygen ion source assembly located at the lower-left corner (Veeco grid-less Mark I ion source), and a substrate assembly located above both the electron beam assembly and the ion source assembly. The substrate assembly has two types of rotations: main platform and satellites (double rotation planetary fixture). The main platform has a radius of approximately 0.34 m and each of the three satellites has a radius of approximately 0.13 m. The main platform rotates at a rate of approximately four revolutions per minute (rpm). Each of the satellites rotates at a rate of approximately 15 rpm. There are three substrate holders in one satellite, with the diameter of each substrate holder being approximately 0.1 m. Thus, each satellite is exposed to the oxygen ion source for an equivalent portion of the time during one rotation of the main platform. This arrangement is slightly different from other conventional ion assisted electron beam deposition processes, where the ion bombardment and vapor condensation simultaneously take place[15]. In the arrangement illustrated in Figure 1, the electron beam (with ~ 7-8 kV and ~ 0.045 A) vaporizes CeO_2 target materials in a crucible (99.9% pure, 3-6 mm pieces, Cerac Item Number C-1065). The vapor thus produced travels towards the substrate assembly and condenses on each of the substrate surfaces evenly, including substrates labeled as 1A, 1B, 2A and 2B. The oxygen ion beam (with ~ 100-170 V and 1.35-1.77 A) strikes the substrate 2A and 2B surfaces with two purposes: "milling" and oxidation. That is, the function of the oxygen ion beam is to "mill" the recently coated cerium oxide layer, and to facilitate full oxidation. While the films on the substrates labeled as 2A and 2B are being "milled," other substrates labeled as 1A and 1B are only receiving vapor condensation. Some time later, the films on the substrates labeled as 1A and 1B will be "milled," and yet the substrates labeled as 2A and 2B are only receiving vapor condensation. Thus, the ion beam only "mills" the films in a periodically pulsed form. The film coating rate for the ion assisted electron beam deposition is typically 0.5 Angstroms per second. The maximum thickness of each film is usually less than 1.5 micrometers. The vacuum chamber shape is nearly cubic, with the length of each side being nearly 0.6 m. The rough pump is a mechanical pump (not shown), with a pumping speed of approximately one cubic meter per minute. The high vacuum pump (cryo-pump) has a diameter of 0.2 meters. The minimum pressure achievable in this vacuum chamber is 22.7 micro-Pa. At the beginning of each deposition, the base pressure of the chamber is 533 micro-Pa or lower. The oxygen flow rate for the ion source is 15-16 micro-standard-cubic meters. Additional oxygen flow is provided to maintain a chamber pressure around 53 milli-Pa during the deposition. The substrate materials include silicon wafers with the orientation of <111> and the diameter of 0.1 m, and low carbon steel plates (1018 Carbon Steel). During deposition, the substrate temperature is maintained at a temperature of approximately 423 Kelvin, which is also the chamber temperature for this system since the temperature is being controlled by a feedback loop for the quartz lamp.

EQUIPMENT UTILIZED TO CHARACTERIZE THIN FILMS

Both EDS and SEM measurements are conducted with one integral system, in which the instrumentation utilized for EDS is the Edax Genesis System, and the equipment for SEM is the FEI Co.'s Quanta 200F. The XRD measurement is carried out with SIEMENS D 500 equipment

(KRISTALLOFLEX 810 X-ray source, cooper radiation). In a typical XRD measurement, the step size for 2-theta is 0.04 degrees, and the dwell time at each step is 9.0 seconds. A normal XRD measurement requires about five hours for one scan, when 2-theta angle varies from 10 degrees to 90 degrees. Raman Spectroscopy is performed with the WiTec Confocal Raman Microscope CRM 200, equipped with a 532.2nm laser[27]. Variable Angle Spectroscopic Ellipsometry measurements are acquired with a monochromator-based VASE instrument from J.A. Woollam Co. VASE data were analyzed with Woollam's WVASE32 software version 3.453. Through-plane electrical impedance of the films is measured using a Solatron 1260 impedance/gain-phase analyzer equipped with a custom-control system. The samples are heated at 5 Kelvin per minute and held at the measurement temperature 10 minutes prior to taking a measurement. At each temperature, impedance is measured using an excitation voltage of 1 V with frequency scanning from 10 MHz to 1 Hz. Equivalent circuit analysis using ZView software enables the extraction of sample resistivity from the impedance spectra, which in turn enables the determination of activation energy. The tensile strength measurement is conducted with Instron 8562 mechanical testing system.

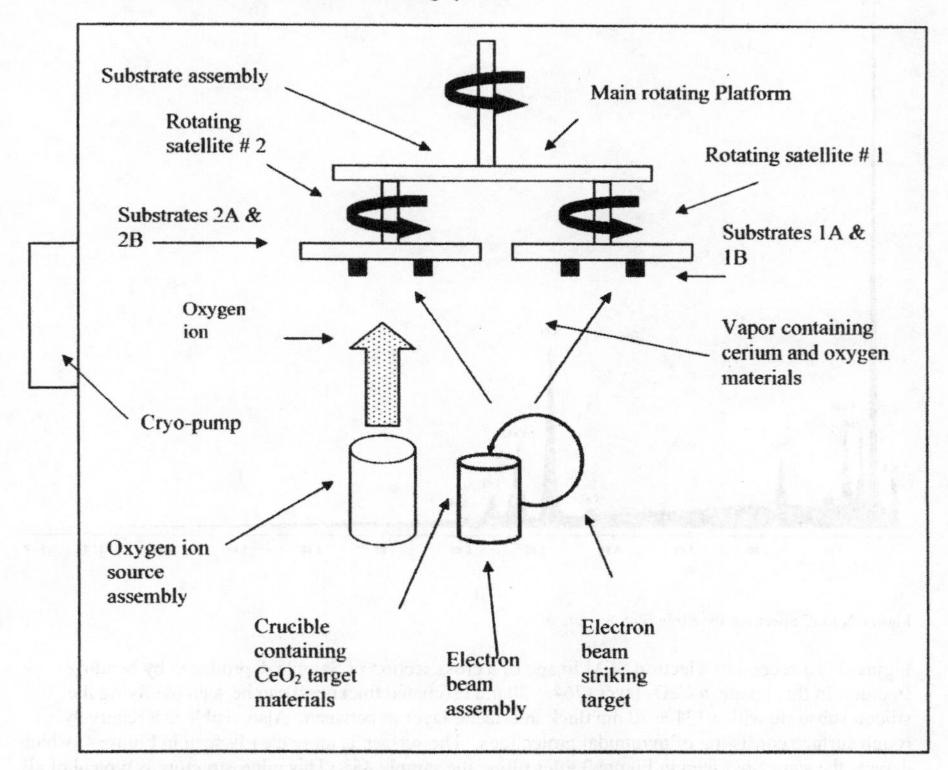

Figure 1. Illustration of ion assisted electron beam deposition system

EXPERIMENTAL RESULTS

Qualitative result of energy-dispersive spectroscopy (EDS) of Sample A is presented in Figure 2. This sample, with a coating thickness of approximately 0.9 micro-meters, is tested with the electron beam normal to the coating surface. The spectrum clearly indicates the presence of cerium and oxygen in significant quantities. The thin film allows penetration of the beam to the substrate, resulting in the small silicon peak that can be observed. The conductive carbon coating applied prior to SEM imaging and EDS is responsible for the carbon peak found in Figure 2.

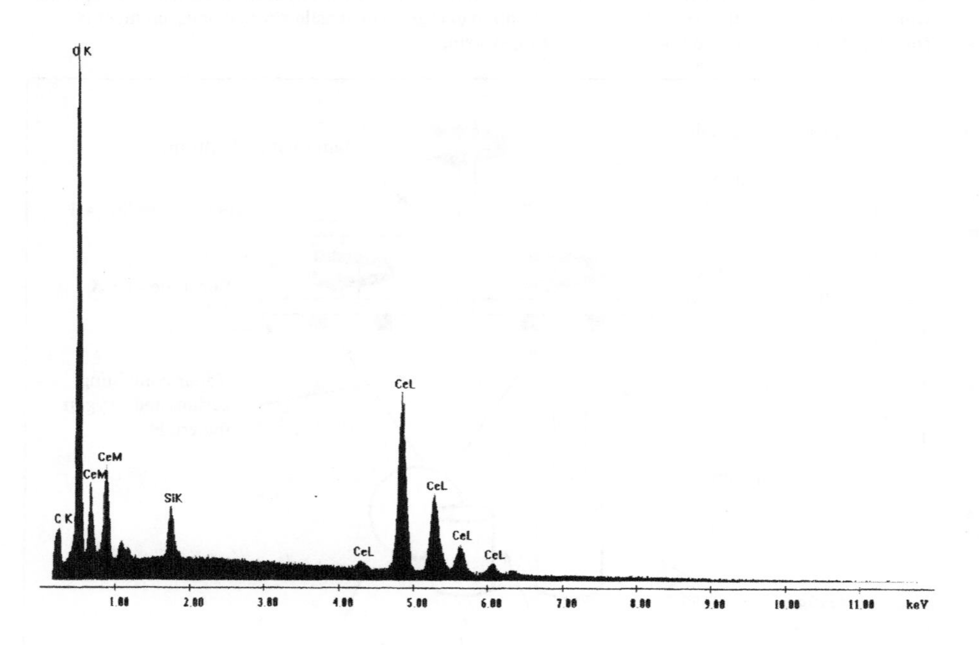

Figure 2. EDS Spectrum for a thin film. Sample A.

Figure 3 is a secondary electron SEM image of a cross section of Sample A produced by bending fracture. In this image, a CeO_2 layer (764 ± 20 nm estimated thickness) can be seen overlying the silicon substrate with a 131 ± 20 nm thick interfacial layer in between. Also visible is a relatively rough surface consisting of pyramidal projections. The surface is more clearly seen in Figure 4, which depicts the same area seen in Figure 3 after tilting the sample 45°. This microstructure is typical of all coating samples produced by this method as supported by Figures 5 through 7. These secondary electron SEM images of Sample B show details of the crystalline morphology and arrangement within the coating. As can be seen in Figures 5 and 6, triangular platelets are the predominant crystalline growth form. The maximum width of the platelets is approximately 220 nm, with average thickness being 20 nm or so. These triangular-shaped platelets are arranged in somewhat regular stacks projecting outward approximately normal to the silicon substrate surface. This imparts a columnar appearance to the fracture surfaces of the coating (see Figure 7). Present on the free surfaces of the triangular platelets and in the inter-columnar spaces are equi-axed crystalline or granular particles averaging 30 ± 5 nm in size (see Figures 5 through 7).

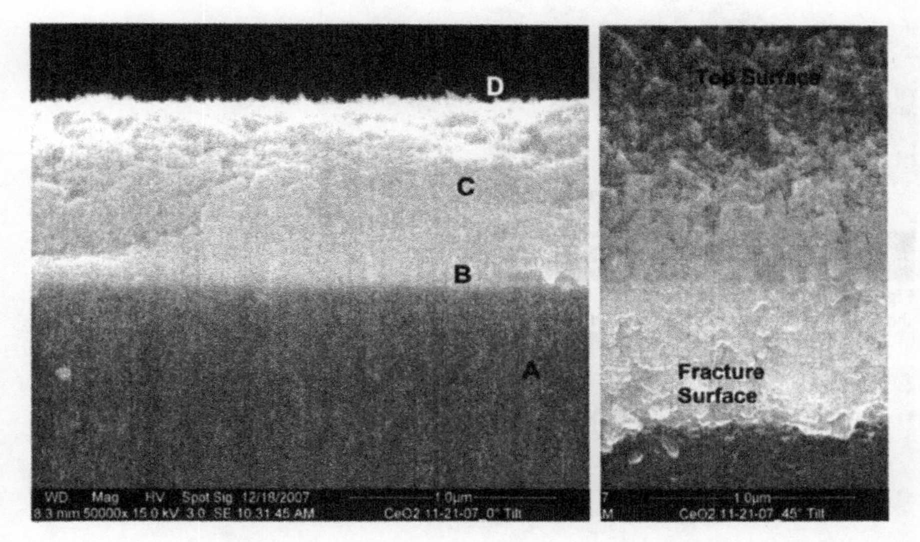

Figure 3. Secondary electron SEM image of cross-section of Sample A showing the silicon wafer substrate (A), interfacial layer (B), and CeO_2 coating (C). Pyramidal projections (D) are also visible on the surface of the coating.

Figure 4. Secondary electron SEM image of area of Sample A shown in Figure 3 rotated 45° to show the surface morphology.

Figure 5. Secondary electron SEM image of Sample B viewed normally to the coating surface showing individual triangular platelets responsible for the surface morphology

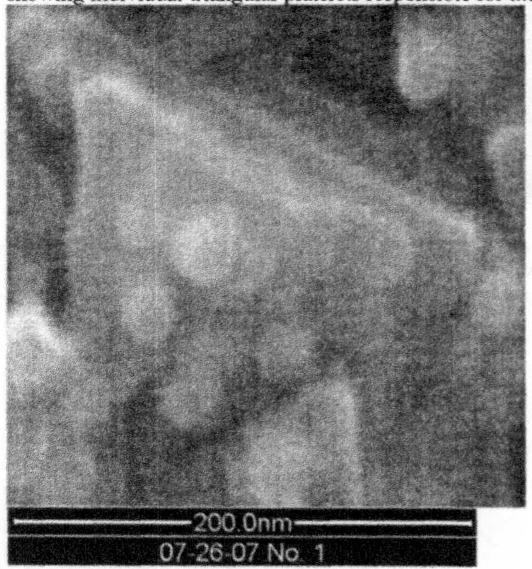

Figure 6. Higher magnification secondary electron image of surface in Figure 5, showing the triangular plate-like crystalline growth

Figure 7. Cross-sectional view of fractured Sample B film, in which layer-stacking is illustrated

In Figure 8, two X-ray Diffraction (XRD) patterns are shown. The top pattern is the XRD pattern for Sample A film, and the bottom pattern is the standard powder diffraction pattern of the CeO_2 powder material (ICDD 00-002-1306). The diffraction pattern of the thin film is closely matched with the standard pattern. However, the number of the peaks for the thin film is less than that of the powder material, which usually indicates the partial alignment of the film. The main peak near 29° of 2-theta angle is <111> peak of CeO_2. Other peaks near 47°, 56°, 77° and 79° correspond to <220>, <311>, <331> and <420> peaks of CeO_2. Notice that the unit cell of CeO_2 is cubic. With the "sheet-stacking" growth, there is a tendency to form a pyramid. Recall that in Figures 5 and 6 some triangular objects are shown. The alignments of CeO_2 are presumably along several axes. Furthermore, based on the FWHM (full width half maximum) of the diffraction peaks, the average crystallite size of CeO_2 is estimated to be 11-18 nm[24-26]. Raman image spectroscopy is utilized to chiefly analyze the Raman spectrum of a thin film sample. The incident laser light impinges upon the top surface of the sample. A Raman spectrum for Sample A is illustrated in Figure 9. There are two peaks between wave-numbers of 300 and 720 cm^{-1}. The peak near 465 cm^{-1} is due to CeO_2 materials[19], and the peak near 520 cm^{-1} is due to silicon[20]. A band-pass filter centered around 465 cm^{-1} is utilized to isolate the CeO_2 signature in a depth probing scan with the Raman equipment. The signature is then colored as bright yellow-orange appearance for visibility as illustrated in Figure 10. The bright area in the middle of the figure is the CeO_2 layer area. The thickness of the CeO_2 layer is estimated to be 765 ± 30 nm, which is close to that illustrated in Figure 3, for the same sample.

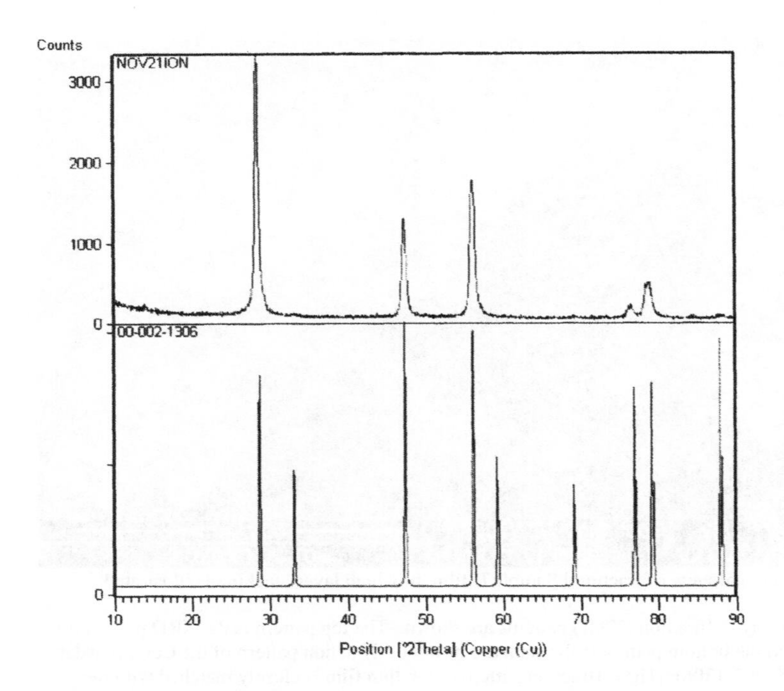

Figure 8. XRD of Sample A film on the top, standard XRD of CeO_2 on the bottom

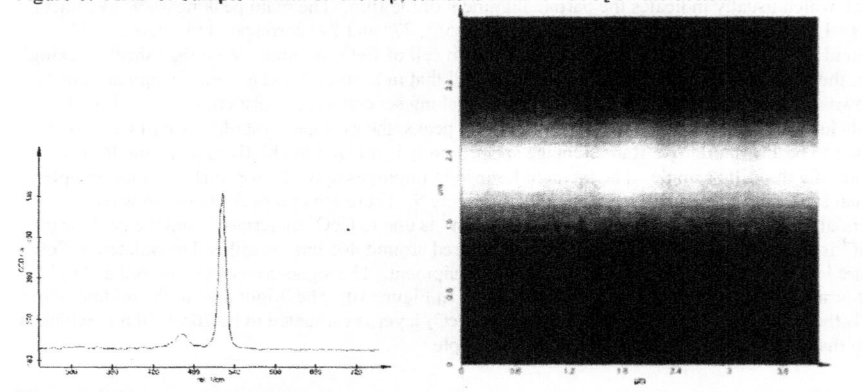

Figure 9. Raman spectrum of Sample A

Figure 10. Raman Imaging Depth-Scan of Sample A

The VASE measurements are analyzed to determine film thickness and refractive index. A single-layer on substrate model does not adequately match the data curves. Fit improvements suggest the presence of three primary microstructural features: anisotropy, grading versus depth, and surface roughness. The refractive index n is a function of wavelength λ, and is described using a two-term Cauchy relationship: $n(\lambda) = A + B/\lambda^2$, where A and B are two parameters[28]. The anisotropy is uniaxial with the difference between the in-plane index and the index normal to the sample surface. No in-plane anisotropy appears, as the microstructural features are randomly oriented within the large area of the measurement spot. The index also decreases toward the surface of the film. Applying a linear gradient where the film is divided into sublayers gives a 16-21% index decrease between the bottom and top of the film. The in-plane index at the top and bottom of the film is shown in Figure 11. The out-of-plane index is lower by about 0.08 from the in-plane values. The final film thickness is 780 ± 10 nm. The surface roughness is modeled using a Bruggeman effective medium approximation to mix the surface index with 50% void. The roughness layer fits to 28 ± 3 nm.

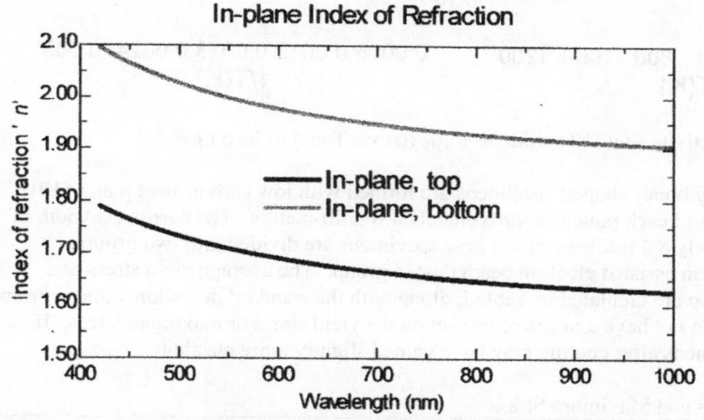

Figure 11. Index of refraction vs. wavelength of the incident light

Figure 12 (a) shows the conductivity of sample A (Figures 2 – 4) as a function of temperature. The through-plane conductivity ranged from ~ 2 x 10^{-9} S/m at 373 K to 6 x 10^{-4} S/m at 1,073 K. For ionic conductors exhibiting a single conduction mechanism, conductivity is expected to depend on temperature as:

$$\sigma = \frac{\sigma_o}{T} e^{-Ea/kT} \tag{1}$$

where σ is conductivity, σ_o is a constant, T is temperature in Kelvin, E_a is activation energy, and k is Boltzmann's constant. Thus, activation energy can be determined from a plot of ln (σT)-vs-1/T, as illustrated in Figure 12 (b). From three repeated measurements, the activation energy of this sample is 0.62 ± 0.08 eV. For another film, Sample C, from four repeated measurements, the activation energy is 0.62 ± 0.10 eV. The activation energy value for these films is comparable to that of the bulk material and some of the thin film materials[1, 21-22].

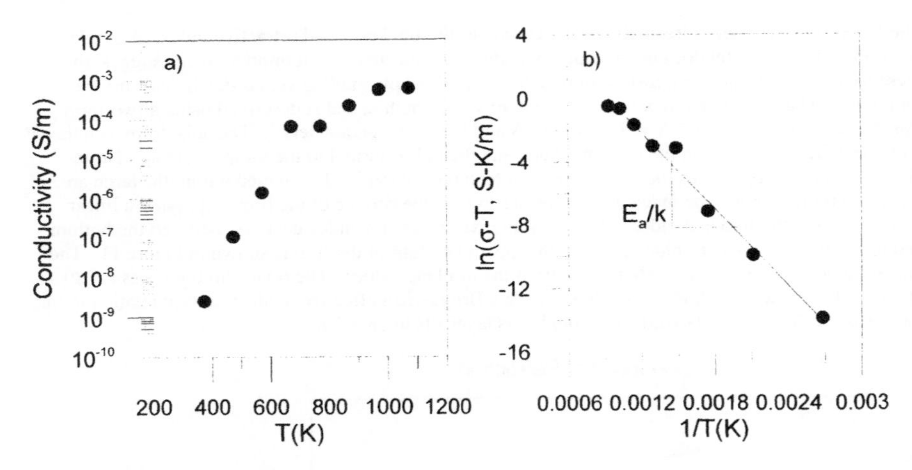

Figure 12. Electrical conductivity of a CeO_2 film: a) $\log_{10}(\sigma)$ vs. T and b) $\ln(\sigma T)$ vs.1/T

Approximately sixty flat "dog bone" shaped specimens are formed with low carbon steel plates (1018 Carbon Steel). The thickness of each plate is approximately 1.6 milli-meters. The narrowest width of each specimen is approximately 8.4 milli-meters. These specimens are divided into two groups: uncoated-control group and ion assisted electron beam coated group. The average yield stress and maximum stress of each group are tabulated in Table I, along with the standard deviation values. From Table I, the coated samples do not have a negative impact on the yield stress or maximum stress. If there is any effect, the specimens after coating may have gained slightly more elasticity.

Table I. Average Yield Stress and Maximum Stress

	Maximum Stress (MPa)	Standard Deviation for Maximum Stress (MPa)	Yield Stress (MPa)	Standard Deviation for Yield Stress (MPa)
Uncoated Control Group	466.6	5.0	401.1	7.3
Ion Assisted Electron Beam Coated Group	468.2	8.1	428.0	11.2

CONCLUSION/DISCUSSION

Based on the experimental results, the as-deposited films are CeO_2 films, with a "sheet-stacking" growth. The crystallite size of CeO_2 is estimated to be 11-18 nm. The film is optically anisotropic; the out-of-plane index values (normal to the sample surface) are lower than the in-plane index values. Both in-plane and out-of-plane index values are a function of the film depth, with the index decreasing towards the surface of the film. The activation energy is 0.62 ± 0.10 eV. Based on the initial DC resistance measurement of several CeO_2 films at the room temperature, the resistivity of the material is estimated to be $3 - 7 \times 10^5\ \Omega$-m. More measurements are currently being conducted to confirm the result.

ACKNOWLEDGMENT

This research is partially supported by NYSTAR-ITCOLLABORATORY and NYSTAR-CACT. Thanks to Prof. R. Giese at SUNY Buffalo for some of the XRD Measurements, to R. Crocker and T. Meehan at VPT for the deposition equipment design and support, to Dr. A. Shamsi of U.S. Department of Energy, National Energy Technology Laboratory (NETL) for the scientific collaboration, to Prof. M. Jackson at RIT and D. Kalapodas at Alfred State for dielectric constant measurements, and to following personnel at Alfred University for the help provided in various aspects of the experiments and data interpretation: S. Zdzieszynski, Dr. W. Mason, F. Williams, J. Amoroso, M. Naylor, J. Thiebaud, Prof. R. DeRosa, Prof. G. McGowan, I. Szabo, S. Molina-Landaverde, J. Ovenstone, and R. Lewis.

REFERENCES

[1] See, for example, Catalysis by Ceria and Related Materials, edited by A. Trovarelli, Imperial College Press, London, 2007.
[2] E. S. Putna, J. Stubenrauch, J. M. Vohs, and R. J. Gorte, Ceria-Based Anodes for the Direct Oxidation of Methane in Solid Oxide Fuel Cells, Langmuir, **11**, 4832-4837 (1995).
[3] B. Zhu, C. Xia, X. Luo, and G. Nikiasson, Transparent Two-Phase Composite Oxide Thin Films with High Conductivity, Thin Solid Films, **385**, 209-214 (2001).
[4] N. Izu, W. Shin, I. Matsubara, and N. Murayama, The effects of the Particle Size and Crystallite Size on the Response Time for Resistive Oxygen Gas Sensor Using Cerium Oxide Thick Film, Sensors and Actuators B: Chemical, **94**, 222-227 (2003).
[5] A. Nazeri, P.P. Trzaskoma-Paulette, and D. Bauer, Synthesis and Properties of Cerium and Titanium Oxide Thin Coatings for Corrosion Protection of 304 Stailess Steel, J. of Sol-Gel Science and technology, **10**, 317-331 (2004).
[6] L. S. Kasten, J. T. Grant, N. Grebasch, N. Voevodin, F. E. Amold, and M. S. Donley, An XPS Study of Cerium Dopants in Sol Gel Coatings for Aluminum 2024-T3, Surface and Coatings Technology, **140**, 11-15(2001).
[7] T. Suzuki, I. Kosacki, H. Anderson, and P. Colomban, Electrical Conductivity and Lattice Defects in Nanocrystalline Cerium Oxide Thin Films, J. Am. Ceram Soc., **82**, 2007-2014 (2001).
[8] S. Guo, H. Arwin, S. Jacobsen, K. Jarrendahl, and U. Helmersson, A spectroscopic Ellipsometry Study of Cerium Dioxide Thin Films Grown on Sapphire by RF Magnetron Sputtering, J. Appl. Phys., **77**, 5369-5376 (1995).
[9] A. Morshed, M. Moussa, S. Bedair, R. Leonard, S. Liu, and N. El-Masry, Violet/blue emission from epitaxial cerium oxide films on silicon substrates, Appl. Phys. Lett., **70**, 1647-1649 (1997).
[10] T. Chaudhuri, S. Phok, and R. Bhattacharya, Pulsed-Laser Deposition of Textured Cerium Oxide Thin Films on Glass Substrate at Room Temperature, Thin Solid Films, **515**, 6971-6974 (2007).

[11] F. Wu, A. Pavloska, D. Smith, R. Culbertson, B. Wilkens, and E. Bauer, Growth and Structure of Epitaxial CeO_2 Films on Yttria-Stabilized ZrO_2, to be published in Thin Solid Films based on the web posting of www.sciencedirect.com, accessed on Dec. 6, 2007.
[12] P. Patsalas, S.Logothetidis, L. Sygellou, and S. Kennou, Structure-dependent electronic properties of nanocrystalline cerium oxide films, Physical Review B. **58**,035104-1-035104-13(2003).
[13] S. Logothetidis, P. Patsalas, E. K. Evangelou, N. Konofaos, I. Tsiaoussis and N. Frangis, Dielectric properties and electronic transitions of porous and nanostructured cerium oxide films, Materials Science and Engineering B, **109**, EMRS 2003, Symposium I, Functional Metal Oxides - Semiconductor Structures, Pages 69-73, 2004.
[14] C. Charitidis, P. Patsalas, and S. Logothetidis, Optical and mechanical performance of nanostructured cerium oxides for applications in optical devices, Journal of Physics: Conference Series 10, 226-229 (2005).
[15] R. P. Netterfield, W. G. Sainty, P.J. Martin, and S. H. Sie, Properties of CeO_2 thin films prepared by oxygen-ion-assisted deposition, Applied Optics, **24**, 2267-72, 1985.
[16] I. Shimizu, M. Kumagai, H. Saito, Y. Setsuhara, Y. Makino, and S. Miyake, Synthesis of CeO_2 films by ion beam assisted deposition, Ion Implantation Technology Proceedings, 1998 International Conference, **2**, 943-46, 1999.
[17] See, for example, N. A. G. Ahmed, An improved ion assisted deposition technology for the 21st century, Surface and Coatings Technology, **71**, 82-7, 1995.
[18] Equipment model is VPT Citation 30, purchased from Vacuum Process Technology, Inc., Plymouth, MA. The control system is a VPT PLC control system.
[19] I. Kosacki, T. Suzuki, H. U. Anderson and P. Colomban, Raman scattering and lattice defects in nanocrystalline CeO2 thin films, Solid State Ionics Volume 149, Issues 1-2, , July 2002, Pages 99-105.
[20] F. Villar, J. Escarre, A. Antony, M. Stella, F. Rojas, J.M. Asensi, J. Bertomeu and J. Andreu, Nanocrystalline silicon thin films on PEN substrates, Thin Solid Films, 516, **5**, Proceedings of the Fourth International Conference on Hot-Wire CVD Cat-CVD Process, Pages 584-587(2008).
[21] http://www.emsl.pnl.gov/new/highlights/200609/index.shtml accessed: Dec. 21, 2007.
[22] C. Mansilla, J. P. Holgado, J. P. Espinos, A. R. Gonzalez-Elipe, and F. Yuberto, Microstructure and transport properties of ceria and samaria doped ceria thin films prepared by EBE-IBAD, Surface & Coatings Technology, **202**, 1256-1261, 2007.
[23] See, for example, M. Alvis, F. De Tomasi, A. Della Patria, M. Di Giulio, E. Masetti, M. R. Perrone, M. L. Protopapa, and A. Tepore, Ion Assistance Effects on Electron Beam Deposited MgF_2 Films, J. of Vacuum Science & Technology A: Vacuum, Surfaces and Films, **20**, 714-720, 2002.
[24] The estimation is based on the Scherrer Equation, with a software package called Jade 8, from Materials Data Incorporated (MDI).
[25] S. J. Chipera, and D.L. Bish, FULLLPAT: A Full Pattern Quantitative Analysis Program for X-ray Powder Diffraction, User's Manual, LANL XRD laboratory, 2001.
[26] Powder Diffraction File database, International Centre for Diffraction Data, Newtown Square, PA, PDF, 2002, 34-0394.
[27] WiTec, Ulm, Germany
[28] See, for example, Handbook of Ellipsometry, edited by H. G. Tompkins, and E. A. Irene, William Andrew Publishing (Norwich, NY, USA) & Springer (Heidelberg, Germany), 2005.

FORMATION OF NANOCRYSTALLINE DIAMOND THIN FILMS ON Ti_3SiC_2 BY HOT FILAMENT CHEMICAL VAPOR DEPOSITION

S. L. Yang, Q. Yang, W. W. Yi, Y. Tang
Department of Mechanical Engineering, University of Saskatchewan, 57 Campus Drive, Saskatoon, SK, Canada S7N 5A9

T. Regier, R. Blyth
Canadian Light Source Inc., University of Saskatchewan, 101 Perimeter Road, Saskatoon, SK, Canada S7N 0X4

Z. M. Sun
National Institute of Advanced Industrial Science and Technology (AIST), Shimoshidami, Moriyama-ku, Nagoya 463-8560, Japan

ABSTRACT

Nucleation and growth of diamond thin films on Ti_3SiC_2 by hot filament (HF) chemical vapor deposition (CVD), under typical microcrystalline diamond (MCD) growth conditions, was investigated. Scanning electron microscopy (SEM), atomic force microscopy (AFM), Raman spectroscopy and synchrotron near edge extended X-ray absorption fine structure spectroscopy (NEXAFS) were used to characterize the synthesized diamond. High density diamond-graphite whiskers were initially formed on the surface of Ti_3SiC_2, which induced a high diamond nucleation density. Consequently, dense smooth nanocrystalline diamond (NCD) thin films were synthesized on Ti_3SiC_2 after 1 h deposition. The synthesized NCD films consist mainly of crystalline diamond with a small amount of graphitic sp^2 carbon. The results indicate that Ti_3SiC_2 has great potentials to be used as substrate materials for NCD diamond thin film deposition.

INTRODUCTION

Diamond, the hardest known natural material, has many other unique properties: it has low coefficient of thermal expansion, is chemically inert, offers low friction and excellent wear resistant, has high thermal conductivity, exhibits negative electron affinity and excellent biocompatibility, and is optically transparent from the ultraviolet to the far infrared[1]. The combination of these properties makes diamond an ideal thin film material for a wide range of applications in optics, semiconductors, microelectronics, biomedical, and manufacturing engineering. CVD techniques have been widely used for diamond thin film synthesis since early 1980s[2-3]. Under the typical CVD diamond synthesis conditions, the diamond crystals in the thin films are typically in micrometer scale size and randomly oriented, and the surface is too rough for many applications. Furthermore, due to the high hardness and excellent chemical stability, the polishing of diamond thin films is rather difficult. Thus, it is desirable to synthesize highly smooth diamond thin films directly. NCD thin films can have very high surface smoothness to meet the surface demand for many applications and have been attracting increasing interests in the last few years.

Nucleation is the first and the most crucial step for NCD thin films formation. Up to now, various non-diamond materials have been used as substrates for CVD diamond thin film growth[4]. However, diamond nucleation on hetero-substrate surfaces without pretreatment is reported to be rather difficult[4]. Recent research results indicated that the feasibility of diamond formation is related to the diffusivity and the solubility of carbon in substrates: the lower the diffusivity and the solubility of carbon, the higher the diamond nucleation density on the substrate[5]. Based on this, one can foresee that diamond might have high nucleation rate and density on carbide hetero-substrates. Recently, we have reported that, by microwave plasma enhanced CVD, adhesive diamond thin films can be synthesized on Ti_3SiC_2 with a high nucleation density and a high growth rate, and NCD thin films can be feasibly synthesized on Ti_3SiC_2 under the typical conditions for MCD formation. These results demonstrated that Ti_3SiC_2 is a promising substrate and interlayer material for diamond nucleation and growth[6-7]. As Ti_3SiC_2 combines properties of both ceramics and metals, the successful synthesis of adhesive diamond thin films on Ti_3SiC_2 may greatly expand the applications for both Ti_3SiC_2 and diamond thin films.

However, the nucleation mechanisms of diamond on Ti_3SiC_2 have not been fully understood up to now. In this work, we investigate the nucleation of NCD thin films on Ti_3SiC_2 by HFCVD, the cheapest and most feasible technique for diamond thin film deposition, under typical MCD growth conditions. It focuses on the early stages of structure development of diamond thin films.

EXPERIMENTAL METHODS

Ti_3SiC_2 slices of 5×10×2 mm were cut by electric-discharge machining from bulk Ti_3SiC_2 prepared by pulse discharge sintering (PDS)[13] and ground with SiC papers down to grits of 600 followed by the ultrasonic cleaning in acetone. The Ti_3SiC_2 samples are composed mainly of Ti_3SiC_2 phase, as indicated by the X-ray diffraction (XRD)[6].

The diamond deposition experiments were performed using a HFCVD system as described earlier[14]. The filament was a coiled tungsten wire of 0.3 mm in diameter and was heated by an ac power supply at a voltage around 30 V and a current around 10 A. A thermocouple was mounted right behind the substrate for the temperature measurement. With a typical distance between filament and substrate of 8 mm, the measured temperature during deposition was about 640 °C. After placing substrates on the substrate holder, the deposition chamber was pumped down to a base pressure of 10^{-3} Pa using a diffusion pump. Then gas mixture of hydrogen (H_2) and methane (CH_4) was introduced into the vacuum chamber using mass flow controllers. The gas flow rate was 0.8 sccm for CH_4 and 50 sccm for H_2. When the working pressure was stabilized at the 500-700 Pa (typically 660 Pa), a current was passed through the tungsten filament coil. These conditions are typical for MCD thin films deposition. The deposition durations ranged from 5 min up to 1 h.

The deposited films were analyzed by scanning electron microcopy (SEM), micro-raman spectroscopy and synchrotron near edge extended X-ray absorption fine structure spectroscopy (NEXAFS). The Raman spectra were obtained using a Renishaw Micro-Raman System 2000 Spectrometer operated at the argon laser wavelength of 514.5 nm. The laser spot size was approximately 2 μm with a power of 20 mW. The NEXAFS experiments were performed at the High Resolution Spherical Grating Monochromator (SGM) beamline of the Canadian Light Source Inc. (CLS), University of Sasktchewan. NEXAFS were recorded in total electron yield (TEY), recorded by monitoring the current generated through photon absorption.

Raman spectroscopy is the primary tool to distinguish between the sp^2 graphite structure and sp^3 diamond structure, this technique suffers from several drawbacks for the quantitative characterization of nanostructured carbon materials. One main drawback is that the difference in the Raman cross sections for sp^2 graphitic features and sp^3 diamond hybridized carbon is large. The Raman cross section for sp^2 carbon can be up to 233 times higher than that for sp^3 diamond when a 514.5 nm wavelength argon ion laser is employed[8], the spectra observed is thus completely dominated by the scattering from the sp^2 carbon when the amount of sp^2 carbon is over 5%[9]. A second shortcoming of the Raman measurement is its dependence on the long range order parameter of the materials. The Raman incident photon wavelength is on the order of microns, which leads to strong crystal size dependence and a critical crystallite size. Alternatively, NEXAFS is sensitive to short range order and local chemical bonding, and has similar cross section to different carbon allotropes, and thus more suitable to differentiate nanostructured sp^3 diamond phase from nano sp^2 graphite[10-12]. In the present study, both Raman spectroscopy and NEXAFS was employed and compared for the chemical structure identification on the deposited films.

RESULTS AND DISCUSSION

After 5 min deposition, a dark layer is visible on the surface of Ti_3SiC_2. Typical SEM morphology (figure 1) shows that this layer is composed of nanometer-scaled whisker-like structures. However, SEM can not provide more detailed information for these nanostructured carbon whiskers due to its limited resolution.

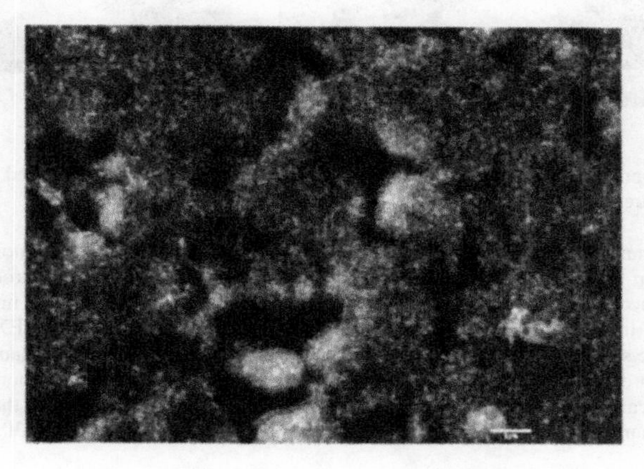

Figure 1 SEM surface morphology of the initial layer formed on Ti_3SiC_2 after 5 min deposition

The structure of these carbon whiskers were further analyzed by AFM, as shown in figure 2. The average length of these whiskers is around 30 nm, and the average width of them is around 5 nm (figure 2a). The higher magnification image, as marked in figure 2b, shows that there are two sub-structures in each whisker: grey (1) and white (2) colored, illustrating that there may be two phases included in each carbon whisker. It is reasonable to assume that graphite and diamond initially formed after short deposition.

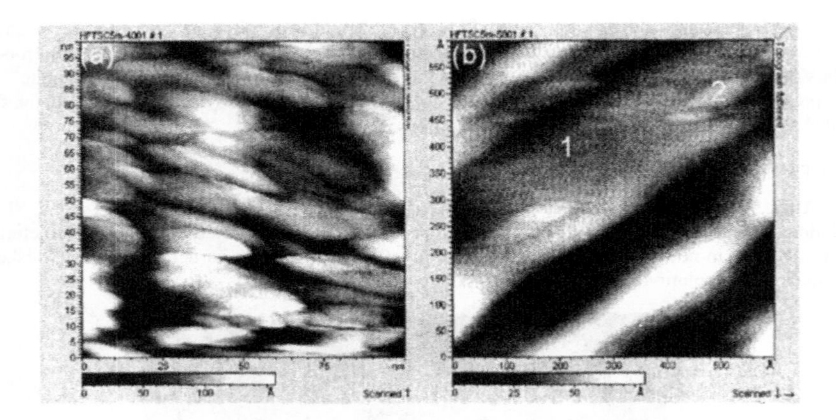

Figure 2 (a) lower, and (b) higher magnification AFM image of the nanostructured carbon whiskers presented in figure 2

The Raman spectra, taken from the layer, are dominated by non-diamond peaks (centered around 1350 cm^{-1} and 1580 cm^{-1}) and the diamond characteristic peak around 1333 cm^{-1} is ambiguous, indicating that Raman spectroscopy with a laser wavelength of 514 nm is not appropriate to examine the initial structure formed after short time deposition. Thus NEXAFS were further conducted on the samples. Figure 3 shows the C K-edge NEXAFS spectrum of nano-structured carbon layer after 5 min deposition. Even though there is a strong peak at 284.5 eV (bonding feature due to the unsaturated carbon in the films[11, 15]), the spectrum is obviously different from that of pure graphite[11] and exhibits a sharp spike at around 289 eV (an excitonic transition) and a dip at 303 eV, the

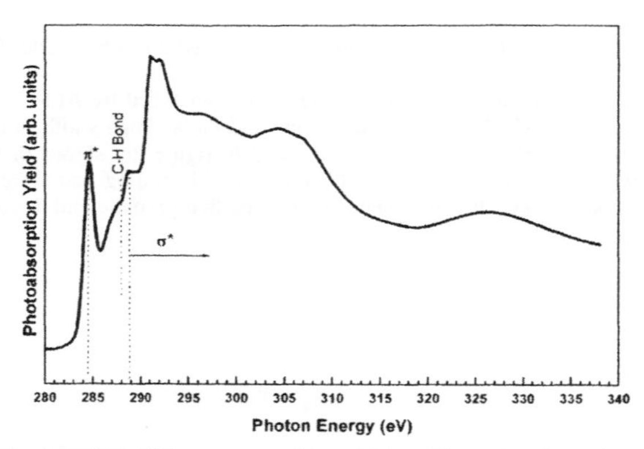

Figure 3 TEY C K-edge NEXAFS spectrum of the whisker-like carbon layer formed after 5 min deposition

characteristic peaks of the pure diamond[10], confirming the diamond structure in the initial layer. The relatively high peak of bonding in the spectrum indicates that this carbon layer consists of relatively high concentration of sp^2 graphitic structure. The weak shoulder at ~288 eV, corresponding to C-H bond in the diamond film[16], can also be seen.

After 1 h deposition, a very dense flat NCD thin films has been formed on the surface of Ti_3SiC_2 (figure 4a). The corresponding NEXAFS spectrum (figure 4b) exhibits a sharp absorption edge at 289 eV and large dip at 303 eV but a weak peak at 285.5 eV, indicating high concentration of diamond phase in the thin film, much higher than the initial layer. Considering that the grain boundaries usually consist of graphitic carbon, the fine NCD thin films are of diamond nature. These results indicate that very high diamond nucleation density and diamond thin film formation rate were achieved on the initially formed nanowhisker-like layer.

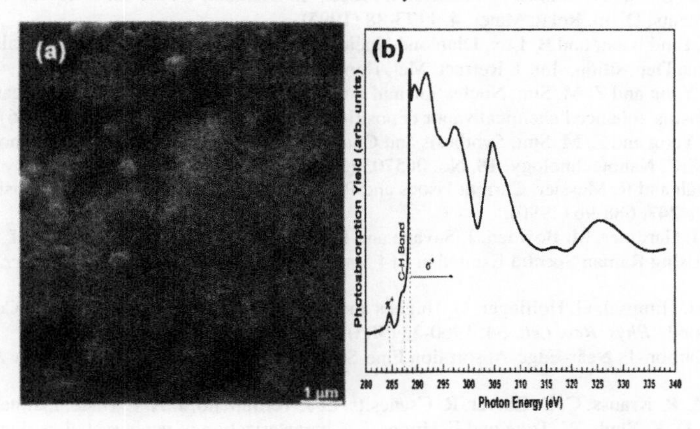

Figure 4 (a) SEM surface morphology and (b) TEY C K-edge NEXAFS spectrum of the nanocrystalline diamond thin film formed on Ti_3SiC_2 after 1 h deposition

CONCLUSION

Nucleation and formation investigation of diamond thin films on Ti_3SiC_2 by HFCVD show that an initial nanowhisker-like diamond-graphite layer was formed on the surface of Ti_3SiC_2, which induces high diamond nucleation density and formation rate. Consequently, dense and smooth nanocrystalline diamond thin films with high purity can be synthesized on the surface of Ti_3SiC_2 in one hour under typical microcrystalline diamond (MCD) growth conditions. The results demonstrate that that Ti_3SiC_2 has great potential for use as substrate materials for NCD diamond thin film deposition and synchrotron based NEXAFS is a powerful tool to distinguish the nanocrystalline sp^3 diamond structure from sp^2 graphite structure.

ACKNOWLEDGMENTS

This work is supported by the Canada Research Chair Program, NSERC and CFI. The diamond deposition in this work was conducted in the Hot Filament Reactor in the Plasma Physics Laboratory, University of Saskatchewan, which was purchased with Prof. Akira Hirose's Canada Foundation for Innovation (CFI) grant. The NEXAFS was performed at Canadian Light Source Inc. (CLS), which is supported by NSERC, National Research Council Canada (NRC), Canadian Institutes of Health Research (CIHR), and the University of Saskatchewan.

REFERENCES

[1] H. O. Pierson, Handbook of Carbon, diamond and fullerenes. Park Ridge, NJ: Noyes; p. 244-77 (1993).
[2] M. Kamo, Y. Sato, S. Matsumoto and N. Setaka, Diamond Synthesis from Gas-Phase in Microwave Plasma J. Cryst. Growth, **62**, 642-4 (1983).
[3] Y. Saito, S. Matsuda and S. Nogita, Synthesis of Diamond by Decomposition of Methane in Microwave Plasma, J. Mater. Sci. Lett. **5**, 565-8 (1986).
[4] H. M. Liu and D. S. Dandy, Studies on Nucleation Process in Diamond CVD – an Overview of Recent Developments, Diam. Relat. Mater. **4**, 1173-88 (1995).
[5] R. Haubner, A. Lindlbauer and B. Lux, Diamond Nucleation and Growth on Refractory Metals using Microwave Plasma Deposition, Int. J. Refract. Met. Hard Mat., **14**, 119-25 (1996).
[6] S. L. Yang, Q. Yang and Z. M. Sun, Nucleation and growth of diamond on titanium silicon carbide by microwave plasma enhanced chemical vapor deposition, J. Cryst. Growth, **294**, 452-8 (2006).
[7] S. L. Yang, Q. Yang and Z. M. Sun, Synthesis and Characterization of Nanocrystalline Diamond Thin Film on Ti_3SiC_2 Nanotechnology, **18**, No. 065703 (2007).
[8] W. A. Yarbrough and R. Messier, Current Issues and Problems in the Chemical Vapor-Deposition of Diamond, *Science* **247**, 688-96 (1990).
[9] S. R. Sails, D. J. Gardiner, M. Bowden, J. Savage and D. Rodway, Monitoring the Quality of Diamond Films Using Raman Spectra Excited at 514.5 nm and 633 nm , *Diamond Relat. Mater.*, **5**, 589 (1996)
[10] J. F. Morar, F. J. Himpsel, G. Hollinger, G. Hughes and J. L. Jordan, Observation of a C-1s Core Exciton in Diamond, *Phys. Rev. Lett.* **54**, 1960-3 (1985)
[11] P. E. Batson, Carbon-1s Near-Edge-Absorption Fine-Structure in Graphite, *Physical Review B*, **48**, 2608-10 (1993).
[12] D. M. Gruen, A. R. Krauss, C. D. Zuiker, R. Csencsits, L. J. Terminello, J. A. Carlisle, I. Jimenez, D. G. J. Sutherland, D. K. Shuh, W. Tong and F. Himpsel, Characterization of nanocrystalline diamond films by core-level photoabsorption, *Appl. Phys. Lett.*, **68**, 1640-2 (1996).
[13] S. L. Yang, Z. M. Sun, Q. Yang and H. Hashimoto, Effect of Al Addition on the Synthesis of Ti_3SiC_2 Bulk Material by Pulse Discharge Sintering Process, J. Euro. Cera. Soc., **27**, 4807-12 (2007).
[14] W. Chen, C. Xiao, Q. Yang, A. Moewes, A. Hirose, The Effect of Bias Polarity on Diamond Deposition by Hot-Filament Chemical Vapor Deposition, Canadian Journal of Physics, **83**, 753-759 (2005).
[15] J. Nithianandam, J. C. Rife, H. Windischmann, Carbon-K Edge Spectroscopy of Internal Interface and Defect States of Chemical Vapor-Deposited Diamond Films, *Appl. Phys. Lett.*, **60**, 135-7 (1992).
[16] M. M. Garcia, I. Jimenez, L. Vazquez, C. Gomez-Aleixandre, J. M. Albella, O. Sanchez, L. J. Terminello and F. J. Himpsel, X-ray absorption spectroscopy and atomic force microscopy study of bias-enhanced nucleation of diamond films, *Appl. Phys. Lett.*, **72**, 2105 (1998)

Thermal Barrier Coating Processing, Development, and Modeling

PROCESS AND EQUIPMENT FOR ADVANCED THERMAL BARRIER COATINGS

Albert Feuerstein, Neil Hitchman, Thomas A. Taylor, Don Lemen
Praxair Surface Technologies, Inc.
Indianapolis, Indiana, USA

ABSTRACT

State of the art advanced thermal barrier coating (TBC) systems for aircraft engine and power generation hot section components consist of EB-PVD applied yttria stabilized zirconia and platinum modified diffusion aluminide bond coating. Thermally-sprayed ceramic coatings are still extensively used for combustors and power generation blades and vanes. This paper highlights the key features of plasma spray and EB-PVD coating processes for TBC. The process and coating characteristics of APS low density and dense vertically cracked (DVC) Zircoat™ TBC as well as EB-PVD coatings are described. The most important bondcoat processes are touched. The major coating cost elements such as material, equipment and processing are explained for the different technologies. New trends in TBC development such as ultra pure Zirconia for improved sintering resistance and low conductivity compositions are addressed.

INTRODUCTION

Thermal barrier coating systems (TBC's) are widely used in modern gas turbine engines to lower the metal surface temperature in combustor and turbine section hardware and to meet increasing demands for greater fuel efficiency, lower NOx emissions, and higher power and thrust. The engine components exposed to the most extreme temperatures are the combustor and the initial rotor blades and nozzle guide vanes of the high pressure turbine. Metal temperature reductions of up to 165°C are possible when TBC's are used in conjunction with external film cooling and internal component air cooling[1]. A diagram of the relative temperature reduction achieved using both TBC and cooling air technologies on hot section hardware is shown in Figure 1.

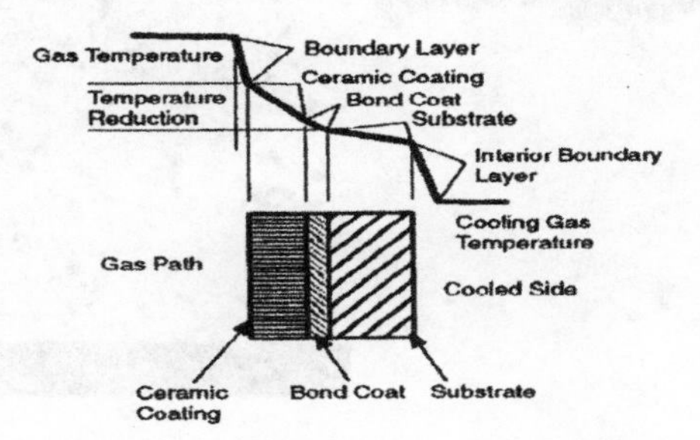

Figure 1. Schematic of a TBC on an air-cooled gas turbine engine component [2]. The thermal barrier acts as a heat flow resistor and is only efficient in air cooled engine components.

A typical thermal barrier coating system consists of two key layers: an oxidation resistant bondcoat such as diffusion aluminide or overlay MCrAlY bond coating, and a plasma sprayed or electron beam evaporated [3] ceramic top layer, typically 7-8 wt.% Y_2O_3-stablilized ZrO_2 (7YSZ), to reduce the heat flux into the component.

Significant effort is going into the development of new ceramic compositions with lower thermal conductivity. Concepts used are advanced multicomponent zirconia (ZrO_2)-based TBC's using an oxide defect clustering design with and materials with a pyrochlore structure [4,5]. A promising innovation consists of doping partially-stabilized YSZ with paired-cluster rare-earth oxides. [6,7,8,9]

An aluminum-enriched bond coat composition is used to provide a slow growing, adherent aluminum oxide film, otherwise know as a thermally grown oxide, or TGO [10]. This alumina scale is an ideal oxygen diffusion barrier, since it has one of the lowest oxygen diffusion rates of all protective oxide films [11]. Bondcoat oxidation is the primary cause of TBC failure [12]. Thermal cycling testing has shown the modes of failure to include the growth of a delamination crack in the zirconia layer just above the TGO and bondcoat, cracking within the TGO, and at the TGO / bondcoat interface [13,14,15].

This paper concentrates on thermal barrier coating systems for selected high temperature applications in a modern aircraft and industrial gas turbine engines. Specifically, the key features of each layer which affect performance, common deposition technologies, and the economic factors are addressed. More recent development at Praxair Surface Technologies, Inc., (PST), towards more sintering resistant TBC are highlighted. The feasibility of low conductivity ceramic compositions as a substitute for 7YSZ as low density plasma TBC and DVC are investigated.

7YSZ CERAMIC THERMAL BARRIER COATINGS

State of the art 7YSZ ceramic top layers are applied either by plasma spray or by electron beam physical vapor deposition (EB-PVD) [2,3]. Alternate thermal spray technologies using liquid precursors show promising results and potential but need still optimization work [16].

Plasma Sprayed Ceramic Thermal Barrier Coatings

A ceramic top layer deposited by air plasma spray (APS) consists of either of two morphologies. A "low density" coating exhibits an even spacing of pores and voids ranging from 20 μm to nano-sized, with sub-critical horizontal micro-cracking between individual splat layers. The coating density usually ranges between 80 and 86 percent of the theoretical value. Figure 2 shows optical and SEM images of a typical low density coating in cross-section.

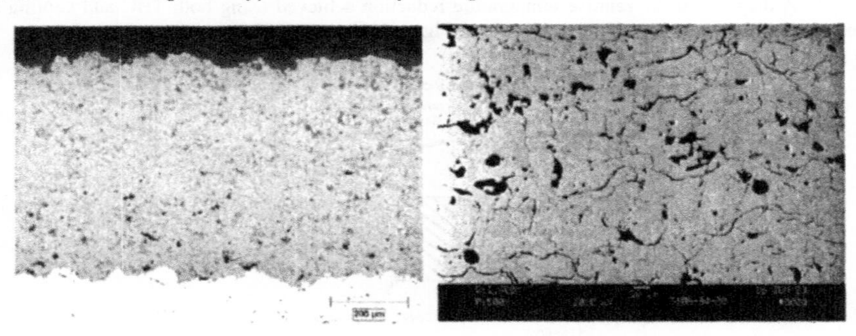

Figure 2: Cross-section of APS low density TBC, containing approximately 15% porosity. Two different magnifications are shown to reveal the "macro" and "micro" structure of the coating.

Alternatively, dense, vertically segmented coatings with improved tolerance of the ceramic layer to the strain caused by the CTE mismatch of ceramic and bondcoat., such as Zircoat™ [17] shown in Figure 3, are successfully used in both aircraft and land-based gas turbine engines [18,19].

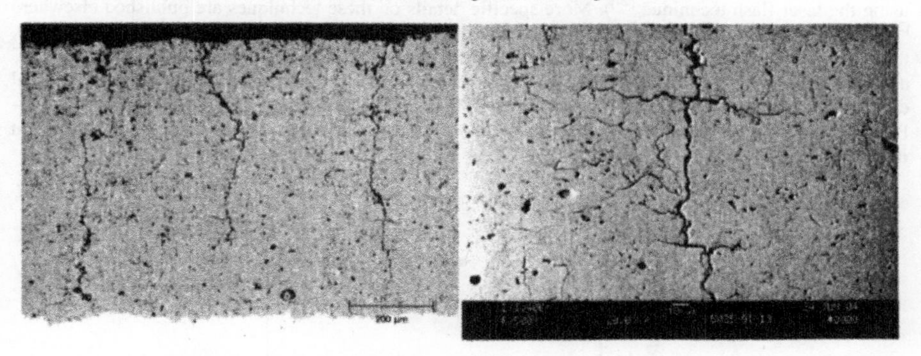

Figure 3: Cross-section of APS Zircoat™ TBC, containing approximately 16-24 cracks per linear centimeter. Two different magnifications are shown to reveal the "macro" and "micro" structure of the coating.

EB-PVD Ceramic Thermal Barrier Coatings

The 7YSZ EB-PVD TBC polished cross-section in Figure 4 shows a plurality of fine columnar grains nucleating on top of an aluminide bond coat. These subsequently increase in size during the vapor deposition process due to competitive growth. The loosely bonded columnar grains provide a high degree of mechanical compliance. However, the lack of large splat boundaries and other features normal to the heat flow direction ensures that 7YSZ EB-PVD TBC's will have relatively higher thermal conductivity values versus their plasma-sprayed counterparts of the same composition [13].

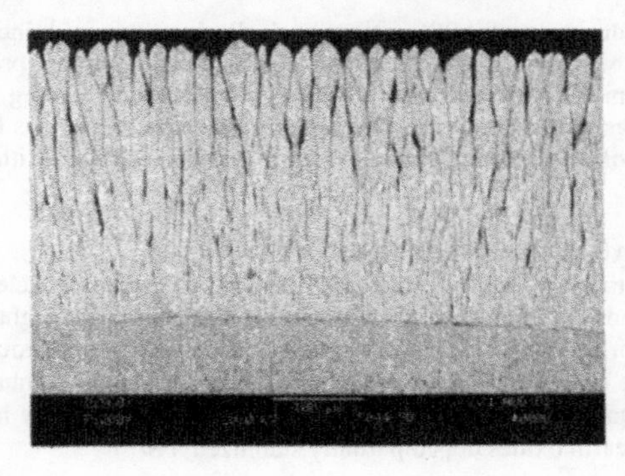

Figure 4: EB-PVD TBC, featuring a plurality of vertical, loosely-bonded columnar grains.

THERMAL CONDUCTIVITY OF APS AND EB-PVD TBC

In order to calculate thermal conductivity, one must know the specific heat of that particular ceramic composition, the density, and the thermal diffusivity as a function of temperature (measured using the laser flash technique [20]). More specific details on these techniques are published elsewhere [21,22]. TBC coatings deposited by APS offer a thermal conductivity significantly lower (0.8-0.9W/mK), than that of fully dense 7YSZ (> 2W/mK). The thermal conductivity of dense vertically cracked TBC and EB-PVD TBC is substantially higher than that of APS coatings, see Figure 5. The nominal densities of plasma sprayed TBC's range from 92% (Zircoat™) to about 85% ("Low Density"). EB-PVD coatings are close to 100% dense in direction of the column growth, however have significant inter-columnar porosity.

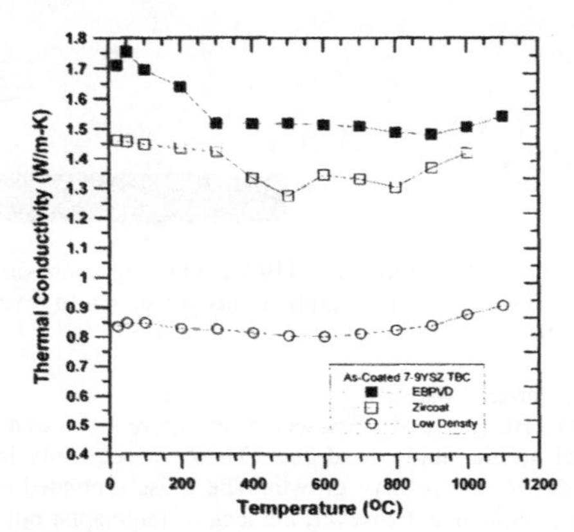

Figure 5: Temperature dependence of thermal conductivity for 7YSZ TBC's.

The thermal conductivity of APS coatings typically increases with increasing pressure, as it is the case in the engine, whereas EB-PVD thermal conductivity is almost pressure independent. The thermal conductivity typically also increases with long time thermal ageing of the coatings (APS as well as EB-PVD coatings) due to sintering effects and phase transformations in 7YSZ. Actually this is a main issue which justifies the today search for new ceramic coatings with low conductivity AND increased stability [23, 24].

ADVANCED THERMAL BARRIER COATINGS AT PST

In the next generation of engines, increased demand for greater efficiency, higher horsepower, and lower NOx emissions will require higher combustion temperatures, lighter weight materials, and less fan air diversion for cooling [4]. At Praxair we are investigating two routes for ceramic coatings with higher temperature capability. We are looking into improvements / enhancements of the current 7YSZ coatings with regard to sintering resistance and also investigate new low conductivity ceramic compositions with rare-earth oxides doped partially stabilized YSZ [6,7,8].

High Purity Sinter Resistant 7YSZ

The improvement comes from using special, highly pure YSZ powders, and spraying with either the standard PST plasma torch or a newly developed plasma torch capable of long standoff coatings of the high density, vertically crack segmented structure. The composition of the new powder compared to the standard present purity material is at least 100 time more pure in the critical alumina and silica analyses. The coatings are called Zircoat-HP™ to make this distinction [25]. Zircoat-HP™ is more thermal shock resistant than even the first version of Zircoat™. It has the same low thermal conductivity, better finishability for smoothness, and has remarkable resistance to density change upon high temperature exposure. The density of standard Zircoat™ after exposures up to 100 hours and up to 1400°C can decrease by nearly 20% due to fine porosity agglomeration. Also, the segmentation cracks begin to close at 1400°C. In contrast, Zircoat-HP™ density does not change (Figure 6) and the segmentation remains after the same exposures [26].

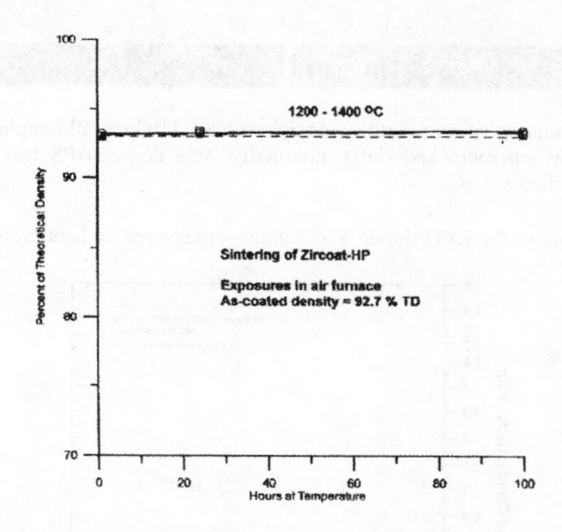

Figure 6: Density of Zircoat-HP™ vesus long term exposure at 1200 and 1400 C is demonstrates the stability of the coating against sintering.

The new long standoff plasma torch developed by PST [27] allows either Zircoat™ or Zircoat-HP™ to be coated on large IGT blades and vanes.

Low Conductivity TBC

In a baseline study we have investigated the feasibility of utilizing lower thermal conductivity coatings of a rare-earth oxide doped YSZ composition in two APS coating morphologies – as low density coating and as dense, vertically macro-cracked (DVM) coating. The APS coatings were generated from spray-dried and sintered powder of a rare-earth oxide doped TBC consisting of Bal. ZrO_2 – 9.23 wt.% Y_2O_3 – 5.12 wt.% GdO_2 – 5.56 wt.% Yb_2O_3 [9]. The micrographs of polished TBC cross-sections shown for the rare-earth oxide doped YSZ samples in Figure 7, exhibit the typical features that are found in APS 7YSZ coatings. The dense vertically macrocracked coatings contain a controlled population of vertical segments between 16 to 24 cracks per linear centimeter.

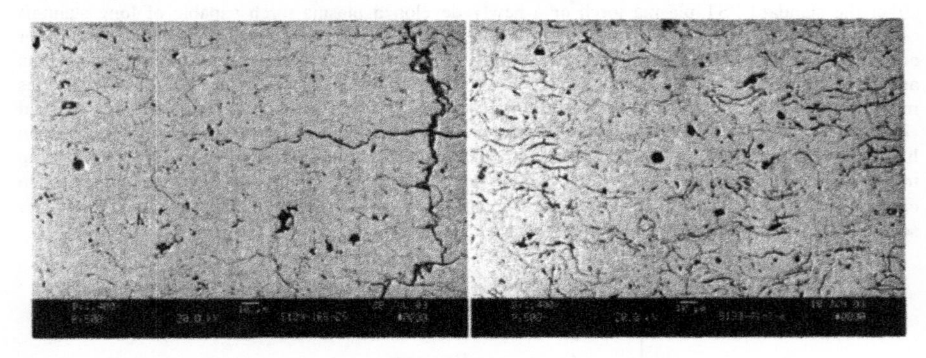

Figure 7: SEM images of rare-earth oxide doped YSZ TBC metallographic cross-sections - APS vertically macrocracked (left), nominally 92% dense. APS low density (right), nominally 85% dense.

The thermal conductivity of the REO-doped YSZ coatings measured on heating is shown in Figure 8.

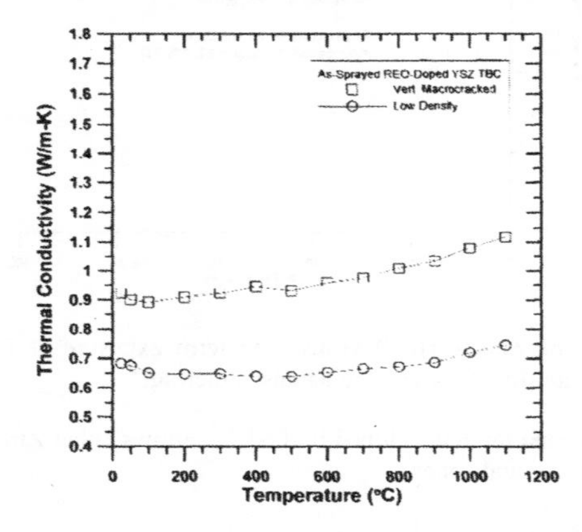

Figure 8: Temperature dependence of thermal conductivity for REO-doped YSZ TBC's

Figure 9 depicts the differences in "net benefit," or percentage reduction in thermal conductivity for each APS coating morphology, resulting from the substitution of rare-earth oxide doped YSZ for 7-8 wt.% YSZ. At room temperature, the dense, vertically segmented TBC's realize the greater "net benefit" from utilizing REO-doped YSZ, with a 38% reduction in thermal conductivity. At 1000°C, this benefit falls to a 24% reduction.

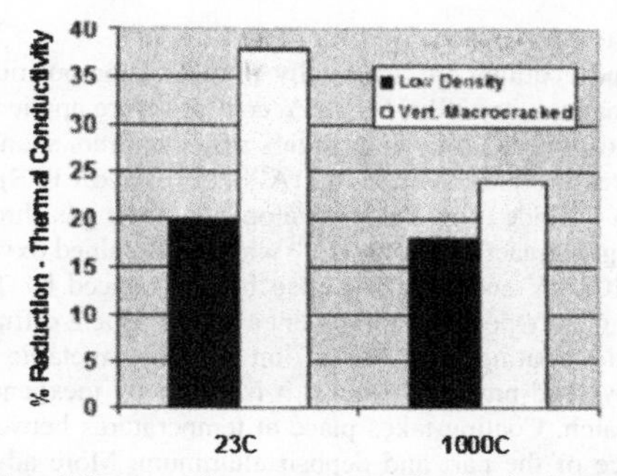

Figure 9: Plot showing differences in percentage reduction in thermal conductivity at 23°C and 1000°C in APS low density and dense vertically macrocracked TBC's when rare-earth doped YSZ is substituted for 7-8 wt. % YSZ.

There is still an issue with the thermal shock resistance of these advanced coating compositions. We observed premature spallation of the coating at the interface to the MCrAlY bondcoat. A solution to this problem is the deposition of a two layer ceramic coating, a first thin 7YSZ adherence layer followed by a second low conductivity thermal barrier layer. This approach combines the excellent thermal stability and shock resistance of the 7YSZ with the better thermal barrier properties of the low conductivity composition. To minimize the stress at the ceramic / ceramic interface of this dual ceramic layer, one must carefully adjust the density of both ceramic layers. Furnace cycle testing showed substantial improvement of the thermal shock resistance of the ceramic / ceramic interface when the density was matched (Figure 10).

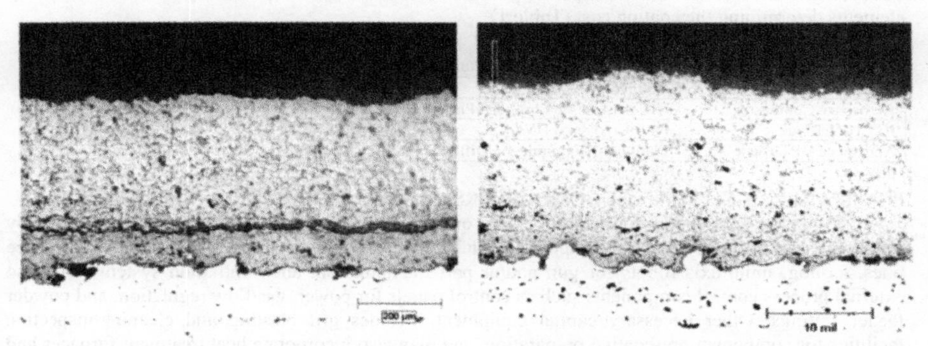

Fig 10: Thermal shock testing of two layer ceramic systems. Bond layer is 7YSZ, top layer is low conductivity TBC.
Left - two layer TBC system – dissimilar density – after 4 FCT cycles.
Right - Improved two layer TBC system –matching density – after 10 FCT cycles.

BONDCOAT

State of the art bond coatings are essentially MCrAlY composition type overlay coatings or diffusion aluminide type coatings. Initially, MCrAlY coatings were applied by electron beam physical vapor deposition. Due to cost considerations, coatings manufacturers soon developed techniques such as air, vacuum, and low-pressure plasma spraying (APS, VPS, and LPPS) and high velocity oxy-fuel (HVOF) deposition. Union Carbide (now PST) developed an inert gas shrouded plasma spray process for MCrAlY and other oxygen –reactive coatings [28] which maintained oxygen pickup to below 0.1 wt. percent. More recently, MCrAlY coatings have also been produced by Tribomet™ [29], a Ni or Co electroplating process with CrAlY powder entrapment and subsequent diffusion heat treatment.

Diffusion aluminide coatings are based on the intermetallic compound β-NiAl. Pack cementation is a commonly used process, because it is relatively inexpensive and capable of coating many small parts in one batch. Coating takes place at temperatures between 800-1000°C. Aluminum halides react on the surface of the part and deposit aluminum. More advanced processes consist of "over the pack" vapor phase aluminizing (VPA) or chemical vapor deposition (CVD). Depending on the activity of the aluminum and the coating temperature one can achieve two coating microstructures [30]. The low activity - high temperature process (1050-1100°C), forms NiAl by outward diffusion of nickel. In the high activity - low temperature process (700-950°C), Ni_2Al_3 and possibly β-NiAl forms by inward diffusion of aluminum. Typically a diffusion heat treatment is applied to form a fully homogeneous β-NiAl layer. The addition of platinum to the diffusion aluminide coating system enhances the diffusion of aluminum [31] into the substrate alloy during the diffusion aluminizing process and drastically improves the oxidation properties of the aluminide coating [32]. Typically 5-10 µm Pt is deposited by electroplating, followed by a diffusion aluminide process.

CURRENT TBC COATING PROCESSES

In the following, we concentrate on three important TBC systems as applied at Praxair Surface Technologies for aero and industrial power gas turbines, their related deposition processes and the key elements determining the coating cost (Table 1).

Table 1: Selected examples of TBC systems for various gas turbine applications.

Example	Bondcoat	Topcoat
Aerospace Combustor	NiCoCrAlY via Shrouded Plasma Spray	Zircoat™ via Air via Plasma Spray
IGT HPT Blade	NiCoCrAlY via HVOF	Low density YSZ via Air Plasma Spray
Aerospace HPT Blade	Platinum-Aluminide via Diffusion	Columnar YSZ via EB-PVD

PLASMA SPRAY COATING OF COMBUSTORS, BLADES, AND VANES

A plasma-spray booth for applying TBCs onto aerospace and IGT combustors consists not only of the physical enclosure and the equipment contained therein, such as torches, gas and power service lines, tooling, multi-axis robots for gun and/or part manipulation, and ventilation systems, but also external process control components such as control panels for power, gas flow regulation, and powder feeder settings. Other necessary capital equipment includes grit blasting and cleaning/inspection facilities for component pre-coating preparation, and may also incorporate heat treatment furnaces and stripping tanks. Both MCrAlY and YSZ ceramic layers are typically applied in the same booth.

For the coating of most annular combustors, a simple two axis manipulator is sufficient to move the plasma torch and rotate the component. Combustors with more complex geometry or with features far off-angle to the torch centerline require the use of robots and turntables or other manipulation equipment with multiple axes of movement.

BOND COATING DEPOSITION PROCESS

Two bond coat deposition techniques are reviewed here, the Shrouded Plasma system and HVOF. Both systems have similarities in their requirements to provide an acceptable coating. In the Shrouded Plasma torch, the effluent is surrounded by an argon shroud to minimize the pickup of oxygen during flight to the component. With this proprietary design, oxygen content in argon-shrouded plasma spray deposited MCrAlY has been shown to contain an order of magnitude lower oxygen content in comparison to conventional air plasma sprayed or HVOF-deposited MCrAlY coatings.

In order to provide optimum adherence for a TBC topcoat, a surface roughness of the MCrAlY bondcoat at least approximately 10 μm Ra is desirable so as to mechanically anchor the layers together. Accordingly, PST has historically applied a two-layer bondcoat system [33] onto combustors and other aerospace and IGT components via gas-shrouded plasma spray [28], in conjunction with vacuum heat treatment. The inner layer is sprayed from a finer cut of powder, and sinters to at least 95 percent theoretical density during heat treatment. This layer has an average roughness of 5 μm. The outer layer of bondcoat is consists of the same composition but of coarser powder, so as to produce a final roughness of at least 10 μm on the surface.

The High Velocity Oxy-Fuel (HVOF) deposition system comprises an oxygen and fuel mixture consisting of either kerosene, propylene, propane, natural gas or hydrogen. The mixture of oxygen and fuel is injected into the combustion chamber and is ignited. The powder is injected internally into the upper stream of the combustion flame, either axially or radially. The ignited gases form a circular flame, which surrounds the required coating powder as it flows through the nozzle. The combustion temperatures can exceed 2800°C depending on gun operating parameters and the fuel type. The flame configuration shapes the powder stream to provide uniform heating and acceleration of the powder particles. Similar to the plasma spray process the selection of gun parameters is based on providing the optimum heating and acceleration of the powder particles by the flame. Typical gas velocities are 1000 to 1200 m/s and can exceed 1500 m/s, depending on which hardware and spray parameters are utilized. To achieve the required bond coat density and surface roughness, MCrAlY applied with this technology can be either a single or dual layer. The schematic in Figure 11 demonstrates the working principle of the Praxair TAFA JP5000 HVOF gun.

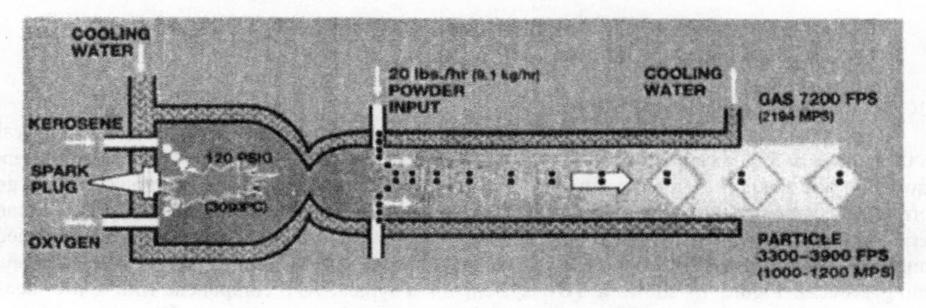

Figure 11: Working principle of the Praxair Tafa JP5000 HVOF gun.

THERMAL BARRIER COATING DEPOSITION PROCESS

The TBC is applied using plasma spray technology, with the resulting coating and its properties significantly influenced by the torch settings. YSZ ceramic powder is injected into the plasma effluent, heated, and accelerated toward the combustor surface. Where this injection takes place, in conjunction with other processing factors, dictates the TBC microstructure. Low density TBC's are best obtained from plasma torches with powder injection ports externally positioned to the exit nozzle. Higher density, vertically segmented TBCs require a more complete treatment of the injected powder. This may be done through external injection with increased torch power, or through internal injection within the torch body.

TBC Systems for Combustors

Figure 12 shows an example of a TBC coating system for a combustor, consisting of a NiCrAlY bond coating and Zircoat ceramic top layer. Generally, the thickness of MCrAlY bond coatings ranges from 125 to 250 µm. The thickness and type of thermal barrier layer is often dependent on the desired morphology, as well as the expected conditions in service. Low density YSZ coatings typically range from 250 to 500 µm, although thicknesses up to 1000 µm are not uncommon in IGT combustors. Dense vertically segmented coatings are generally thicker, due to their higher thermal conductivity. Thicknesses above 500 µm are often recommended for good thermal insulation.

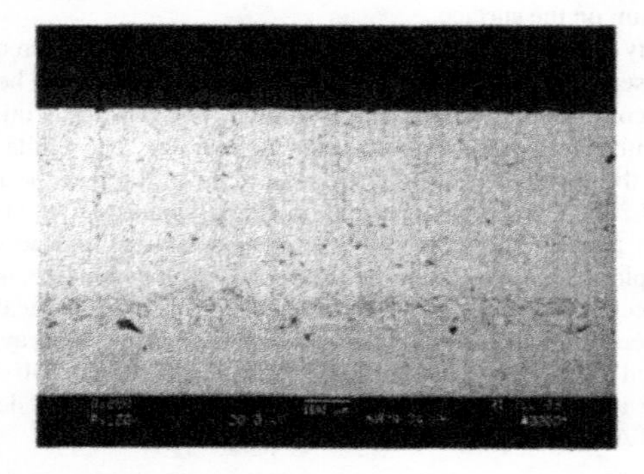

Figure 12: Combustor TBC system consisting of Zircoat™ and NiCrAlY bondcoat.
Total TBC thickness appr. 500 µm.

TBC Systems for IGT Blades and Vanes.

The total coating thickness for IGT components ranges between 400 to 500 µm. Typically, HVOF is used at PST to apply the bond coating due to the large part size and long stand-off distances required during coating. A low density TBC topcoat provides maximum thermal protection and good thermal shock resistance over extended time at high operation temperatures. Plasma-sprayed ceramic thermal barriers are almost always employed. More recently, the Tribomet® process has been applied to complex 3D airfoil geometries which utilize the non line of sight features of the powder entrapment plating process. Figure 13 shows a TBC system for a typical IGT component with a low density thermal barrier coating on a HVOF bondcoat and the same system on a Tribomet® bondcoat. Figure 14 shows a large industrial vane receiving a HVOF bond coating.

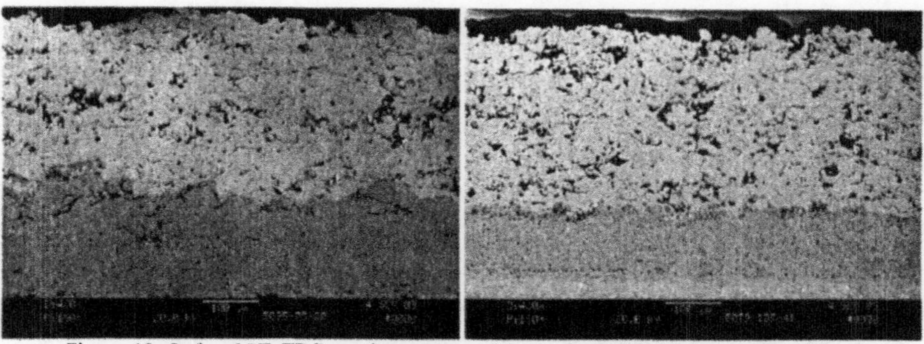

Figure 13: Left - IGT TBC coating system with two-layer HVOF bondcoat and low density TBC. Right - Low density TBC on Tribomet® NiCoCrAlY

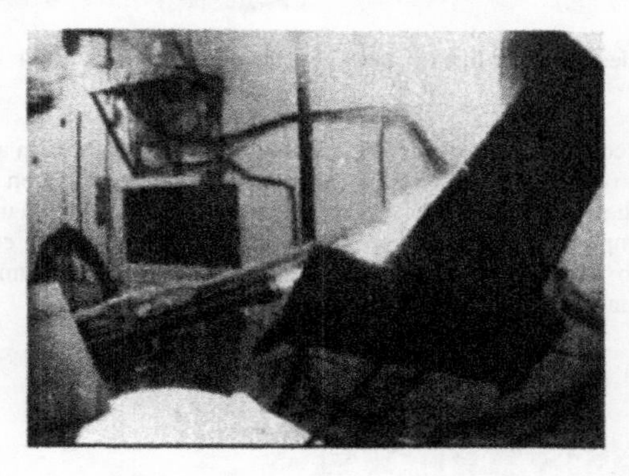

Figure 14: HVOF torch coating the T/E of a large IGT vane

PLATINUM ALUMINIDE / EB-PVD TBC SYSTEM

The platinum aluminide / EB-PVD TBC system can be applied to a variety of modern aircraft and IGT parts. The only limitation is whether the engine hardware can fit into and be easily handled within the various processing chambers. This TBC system requires three separate coating processes: electroplating, aluminizing, and EB-PVD.

Platinum Aluminide Deposition.

Platinum plating and diffusion aluminide are basically batch processes. Depending on the part size, ten to one hundred parts are processed simultaneously. Parts are cleaned in a vapor degreaser; grit blasted, and weighed prior to platinum electroplating (Figure 15a). The electroplated platinum is then diffused in a vacuum furnace prior to the aluminum diffusion. Platinum plated parts are then loaded in a retort containing the aluminum donor and activator for vapor phase aluminizing (Figure 15b). Areas of parts not receiving coating, such as the roots of turbine blades, are masked by nickel paste or tape. Retorts are then placed in the hot zone of a furnace.

Figure 15: Pt Electroplating line (a).and VPA Aluminizing Furnace (b)

EB-PVD TBC

EB-PVD TBC coatings are produced by vacuum deposition of YSZ in a reactive atmosphere at elevated temperatures (app. 1000°C). 7YSZ ingot is evaporated by electron beams in vacuum and deposited onto the preheated parts. By a combination of rotation and tilting, a uniform coating over the airfoil surface is accomplished. To compensate for some oxygen loss during coating, a minor amount of oxygen is added to the process. Typical deposition speeds are 2 - 6 μm per minute. A highly efficient EB-PVD production coater is shown in Figure 16.

Figure 16: EB-PVD TBC 4-chamber production coater.

COST STRUCTURE FOR SELECTED EXAMPLES

The cost factors can be grouped into three cost categories:

A- **One time cost factors** such as application development, tooling development and design, process qualification and approval, production process documentation.

B- **Direct coating related cost factors** such as direct materials (powder, ingot, plating salt, aluminizing donor alloy), auxiliary materials (grit, cleaning agent, gloves); labor for incoming inspection, cleaning and surface preparation, labor for fixturing and masking, post coating processing equipment amortization, energy, process gases.

C- **Indirect cost factors** (materials and services) such as material preparation and recycling, preventive maintenance and spare parts, strip and rework in case of nonconformance, tooling cleaning and rework, tooling replacement, packing and shipping.

Thus, the cost to coat an actual part is very much dependant on the part volume and the production life cycle of a part. With the variability in the cost elements, it can be misleading to state actual coating costs with precision. For each of the coating example given above, it is more useful to state a typical range of costs related to various processing factors. Figure 17 exhibits the main cost groups B and C, as previously mentioned. For this cost breakdown, the following assumptions hold:

- Material costs between \$30 and \$70 per kg for ceramic powder and ingot, and \$50-100 per kg for MCrAlY powder. The price for 1 oz Pt is above \$1000.
- Energy costs are essentially site related, and are based on US averages.
- Gas prices of approximately \$0.1-0.2 per kg for CO2, \$0.5 per m3 for argon and \$2 per liter for kerosene.
- For the investment, an APS plasma cell for approximately \$0.4-0.8 million was considered. Adding an automated parts handling system can add another \$0.5 to 0.8 million. A PtAl diffusion coating facility consisting of a platinum plating line, vacuum heat treat furnaces, and VPA diffusion furnaces costs between \$3 million and \$6 million. Equipment cost is highest for EB-PVD. Depending on the machine capacity, an EB-PVD TBC facility costs between \$15-30 million. An equipment amortization scheme of 10 years depreciation with a 10% annual interest rate is assumed.

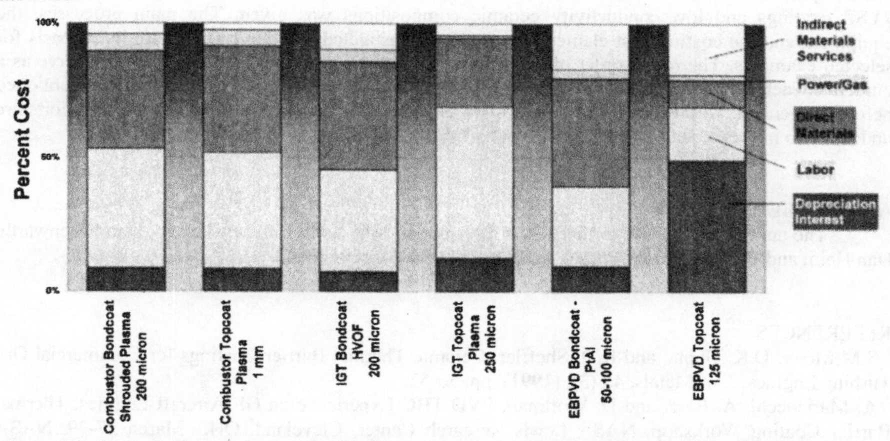

Figure 17: Comparative cost structure for thermal barrier topcoat of selected coating applications.

Discussion of Cost Elements

The cost of the thermal spray process, is essentially driven by the material (powder) cost and labor cost. With APS TBC application, the relatively low deposition rates and comparatively low material expense, labor becomes the most significant portion of the cost. For IGT parts, dependent on the blade or vane size, costs can range between several hundred and several thousand dollar per part. The material cost aspect is even more extreme in the case of platinum aluminide, where platinum can count for several 10$ per part. In the case of EB-PVD with the substantial equipment cost, the equipment depreciation calls for close to 50% of the overall production cost. For EB-PVD TBC only a guideline for the coating cost is for $120-200 per part. For larger aero parts with thicker coating requirements such as nozzle guide vanes, where only 4-6 parts can be coated in one batch, this cost easily triples.

The relative contribution of these cost factors also provides guidance for application development for process and product optimization. Processes with a high percentage of material cost demand improvement of the material utilization. In the case of thermal spray processes, optimized torch parameters can increase the material deposition efficiency substantially and need to be evaluated for each new application. In the case of platinum electroplating, the application development tends towards improving the uniformity and reproducibility of the platinum distribution. This allows adjusting the overall thickness in such a way that the specified thickness requirement is met even at critical locations without having to deposit an excessive thickness on uncritical locations.

Equipment amortization cost dictates the utilization of the facility. In the case of thermal spray processes, where the labor and material cost are the major cost factors, the equipment amortization is moderate; the equipment can be utilized in a one shift or two shift operations. In the case of expensive capital equipment such as EB-PVD and platinum aluminide, 3 shift operation is mandatory. In the case of EB-PVD, development tends toward effective utilization of the expensive capital equipment.

SUMMARY AND CONCLUSION

The key features of state of the art processes for the deposition of TBC systems were outlined and compared. An outlook into PST's new TBC developments such as sintering resistant high purity 7YSZ coatings and low conductivity ceramic compositions was given. The main processes, the equipment and the coating cost elements have been investigated and compared on an index basis for selected examples. The rough order of magnitude cost figures given for selected examples serve as a guideline; each application needs its own detailed evaluation considering all the elements as mentioned before. In general, all presented technologies are considered to be mature. Still, substantial efforts are underway to improve quality and reduce cost by Lean Manufacturing and 6 Sigma programs.

ACKNOWLEDGEMENTS

The authors would like to thank Ann Bolcavage, Adil Ashary, James Knapp, Dan Fillenwarth, Dan Helm and Daming Wang for their input and valuable discussions.

REFERENCES

[1] S.M Meier, D.K. Gupta, and K.D. Sheffler, Ceramic Thermal Barrier Coatings for Commercial Gas Turbine Engines, J. of Metals, 43 (3), (1991), pp. 50-53.

[2] A. Maricocchi, A. Barz, and D. Wortman, PVD TBC Experience on GE Aircraft Engines, Thermal Barrier Coating Workshop, NASA Lewis Research Center, Cleveland, OH, March 27-29, NASA Conference Publication 3312, (1995), pp. 79-90

[3] D.V. Rigney, R. Viguie, D.J. Wortman, D.W. Skelly, PVD Thermal Barrier Coating Applications and Process Development for Aircraft Engines, Journal of Thermal Spray Technology Vol. 6 (2) June 1997 p. 167
[4] D. Zhu, J.A Nesbitt, C.A. Barrett, T.R. McCue, R.A.Miller, Furnace cyclic oxidation behavior of multicomponent low conductivity thermal barrier coatings, Journal of Thermal Spray Technology, Volume 13, Number 1, March 2004, pp. 84-92
[5] D. Stöver; G. Pracht, H. Lehmann, M. Dietrich, J-E. Döring, R.Vaßen , New material concepts for the next generation of plasma-sprayed thermal barrier coatings, Journal of Thermal Spray Technology, Volume 13, Number 1, March 2004, pp. 76-83
[6] J.R. Nicholls, K.J. Lawson, A. Johnstone, and D.S. Rickerby, "Methods to Reduce the Thermal Conductivity of EB-PVD TBCs," Surf. Coat. Technol., 151/152, (2002), pp. 383-391.
[7] D. Zhu and R.A. Miller, "Thermal Conductivity and Sintering Behavior of Advanced Thermal Barrier Coatings," NASA/TM 2002-211481, (2002).
[8] D. Zhu and R.A. Miller, "Low Conductivity and Sintering-Resistant Thermal Barrier Coatings," US Patent 6,812,176 B1, issued Nov. 2, 2004.
[9] D. Zhu and R.A. Miller, "Low Conductivity and Sintering-Resistant Thermal Barrier Coatings," US Patent 6,812,176 B1, issued Nov. 2, 2004.
[10] F.H. Stott, Elevated Temperature Coatings, Science and Technology 11,The Minerals, Metals and Materials Society 1996, pp. 151-161
[11] The Oxide Handbook, Editor G.V Samsonov, IFI / Plenum, 1982
[12] P.K. Wright, A.G. Evans, Mechanisms governing the performance of thermal barrier coatings , Current Opinions in Solid State and Materials Science 4, (1999), pp. 255-265
[13] S. Alperine, M. Derrien, Y. Jaslier, R. Mevrel , Thermal Barrier Coatings – The Thermal Conductivity Challenge, AGARD Report 823 "Thermal Barrier Coatings", 15-16 October 1997, pp. 1.1 -1.10
[14] V. Teixeira, M. Andritschky, H. Gruhn, W. Maliener, H.P. Buchkremer, D., Stoever, Failure of Physically Vapor Deposition / Plasma-Sprayed Thermal Barrier Coatings During ThermalCycling, Journal of Thermal Spray Technology, Vol. 9(2) June 2000 – pp. 191-197
[15] J.A. Haynes, M.K. Ferber, W.D. Porter, Thermal cycling behavior of plasma-sprayed thermal barrier coatings with various MCrAlX bond coats, Journal of Thermal Spray Technology, Volume 9, Number 1, March 2000, pp. 38-48(11)
[16] E.H. Jordan, L. Xie, M. Gell, N.P. Padture, B. Cetegen, A. Ozturk, J. Roth, T.D. Xiao, P.E.C. Bryant, Superior thermal barrier coatings using solution precursor plasma spray, Journal of Thermal Spray Technology, Volume 13, Number 1, March 2004, pp. 57-65(9)
[17] Thomas A. Taylor, US Patent 5,073,433, Dec 17, 1991
[18] T.A. Taylor, D.L. Appleby, A.E. Weatherill, and J. Griffiths, Surf. Coat. Technol., Plasma Sprayed Yttria-Stabilized Zirconia Coatings: Structure-Property Relationships, Surf. Coat. Technol., 43/44, (1990), pp. 470-480.
[19] T. A. Taylor, Dense Vertically Segmented Thermally Sprayed YSZ for TBC and other High Temperature Applications, Proceedings 2nd International Surface Engineering Congress, ASM International, Sept. 2003, Indianapolis.
[20] A. Feuerstein, A. Bolcavage, Thermal Conductivity of Plasma and EB-PVD Thermal Barrier Coatings, Proceedings ASM International Surface Engineering Congress, 2004, Orlando, Florida
[21] R.E. Taylor, Thermal Transport Property and and Contact Conductance Measurements of Coatings and Thin Films, Inter. Journal of Thermophysics, 19 (3), (1998), pp. 931-940.

[22] W. Chi, S Sampath, H.Wang, Ambient and High-Temperature Thermal Conductivity of Thermal Sprayed Coatings, Journal of Thermal Spray Technology, Volume 15, Number 4, December 2006, pp. 773-778
[23] Cernuschi et al., Studies of the sintering kinetics of thick TBC's by thermal diffusivity measurements, J. Europ. Ceram. Soc., 25 (2005), pp. 393-400
[24] Flores Renteria et al., Effect of morphology on thermal conductivity of EB-PVD PYSZ TBC's, Surf. Coat. Technol., 201 (2006), pp. 2611-20
[25] Thomas Taylor, US Patent 5,073,433 and others pending.
[26] Thomas Taylor, Neil Hitchman and Albert Feuerstein, Zircoat-HPTM, A New High Purity Segmented YSZ Coating, ICMCTF Session A2, San Diego, May 2007.
[27] Thomas Taylor and John Jackson, US Patent 7,045,172.
[28] Union Carbide Patent US 3,470,347, Sep 30, 1969
[29] Foster et al, US patent Nos. 5,558,758, Sep 24, 1996 and 5,824,205, Oct 20, 1998
[30] F.S Petit, G.W. Goward, Oxidation – Corrosion-Erosion Mechanisms of Environmental degradation of High Temperature Materials, in Coatings for High Temperature Processes, Editor E. Lang, Applied Science Publishers, 1985
[31] R. Bouchet, R. Mevrel, Influence of platinum and palladium on diffusion in beta-NiAl phase, Defect Diffusion Forum, 237-240 (2005), pp. 238-245
[32] D.K Das, Vakil Singh, S.V. Joshi, The yclic Oxidation Performance of Aluminide and Pt-Aluminide Coatings on Cast i-Based Superalloy CM-247, JOM-e, 52 (1), 2000
[33] Weatherly and Tucker, Union Carbide, U.S. Patent 4,095,003, Jun 13, 1978

CORROSION RESISTANT THERMAL BARRIER COATING MATERIALS FOR INDUSTRIAL GAS TURBINE APPLICATIONS

Michael D. Hill, Ph.D. and Davin P. Phelps.
Trans-Tech Inc.
Adamstown, MD 21710 USA

Douglas E. Wolfe, Ph.D.
Assist Professor, Materials Science and Engineering Department
The Pennsylvania State University
University Park, Pa 16802 USA

ABSTRACT

Thermal Barrier Coatings are ceramic materials that are deposited on metal turbine blades in aircraft engines or industrial gas turbines which allow these engines to operate at higher temperatures. These coatings protect the underlying metal superalloy from creep, oxidation and/or localized melting by serving as an insulating barrier to protect the metal from the hot gases in the engine core. While for aircraft engines, pure refined fuels are used, it is desirable for industrial gas turbine applications that expensive refining operations be minimized. However, acidic impurities such as sulfur and vanadium are common in these "dirty" fuels and will attack the thermal barrier coating causing reduced coating lifetimes and in the worse case catastrophic failure due to spallation of the coating. The industry standard coating material is stabilized zirconia with seven weight percent yttria stabilized zirconia being the most common. When used in industrial gas turbines, the vanadium oxide impurities react with the tetragonal zirconia phase causing undesirable phase transformations. Among these transformations is that from tetragonal to monoclinic zirconia. This transformation is accompanied by a volume expansion which serves to tear apart the coating reducing the coating lifetime. Indium oxide is an alternative stabilizing agent which does not react readily with vanadium oxide. Unfortunately, indium oxide is very volatile and does not readily stabilize zirconia, making it difficult to incorporate the indium into the coating. However, by pre-reacting the indium oxide with samarium oxide or gadolinium oxide to form a stable perovskite ($GdInO_3$ or $SmInO_3$) the indium oxide volatilization is prevented allowing the indium oxide incorporation into the coating. Comparison of EDX data from evaporated coatings containing solely indium oxide and those containing $GdInO_3$ are presented and show that the indium is present in greater quantities in those coatings containing the additional stabilizer. Corrosion tests by reaction with vanadium pentoxide were performed to determine the reaction sequence and to optimize the chemical composition of the coating material. Lastly, select x-ray diffraction phase analysis will be presented.

INTRODUCTION

Thermal Barrier Coatings are ceramic materials that are deposited on metal turbine blades in aircraft engines or industrial gas turbines which allow these engines to operate at higher temperatures. These coatings protect the underlying metal superalloy from creep, oxidation and/or localized melting by serving as an insulating barrier to protect the metal from the hot gases in the engine core.

Several impurities common in fuels have been identified and associated with corrosion in EB-PVD coatings. These impurities include sodium, sulfur, phosphorus and especially vanadium. These impurities react with conventional YSZ turbine blade coatings, severely limiting the coating lifetime.

Therefore, it is of great interest to develop alternative materials that react less readily with fuel contaminants and therefore increase the operating lifetime of the coating.

Standard 8YSZ EB-PVD coatings contain 8-weight percent yttria and crystallize in the metastable t' phase that is derived from a martensitic distortion of the "stabilized" cubic fluorite structure of zirconia. This rapidly cooled t' structure is the most desirable of all of the possible polymorphs in the yttria-zirconia system for TBC applications. Jones[1] described several mechanisms of chemical attack on 8YSZ coatings. These include chemical reaction, mineralization, bond coat corrosion and physical damage due to molten salt penetration. Of the four, only the first two mechanisms will be featured in this discussion.

Acidic species such as SO_3 and V_2O_5 have been shown to react with the yttria stabilizing the t' phase, destabilizing the Y_2O_3-ZrO_2 by extraction of the Y_2O_3. Of these, V_2O_5 has been determined to be the worst offender. Hamilton[2] and Susnitsky[3] have studied the reaction mechanism in detail. The reaction:

$$Zr_{1-x}Y_xO_{2-.5x}\,(t') + yV_2O_5 \rightarrow 2(1-y)\,ZrO_2\,(\text{monoclinic}) + 2y\,YVO_4$$

is especially deleterious to the TBC integrity. The vanadium has been shown to leach the yttria out of the zirconia leaving the yttria deficient monoclinic phase of zirconia remaining. The large volume expansion (7%) caused by this transformation leads to the TBC spalling therefore exposing the bond coat to further chemical attack.

Mineralization, on the other hand, describes a catalytic process by which a metastable phase (in this case, the t' phase) is broken into its stable phase assemblages by a catalyst or mineralizer. For example, ceria stabilized zirconia was investigated as a corrosion resistant coating due to the fact that ceria does not react with vanadium pentoxide.

$$Zr_{1-x}Ce_xO_{2-.5x}\,(t') + yV_2O_5 \rightarrow (1-x)ZrO_2\,(\text{monoclinic}) + xCeO_2 + yV_2O_5$$

However, vanadium does act as a mineralizer, destabilizing the t' phase without reacting to form the vanadate.

<u>Alternate stabilizers for zirconia</u>: A large number of cationic species act to stabilize the cubic and t' phases of zirconia. Therefore, one strategy toward finding corrosion resistant coatings was to find a stabilizer that is resistant to chemical attack by vanadium pentoxide. As mentioned above, ceria was investigated but found to be subject to a mineralization reaction[4]. Previous work at NRL[1] focused on studying acidic stabilizers to zirconia since basic stabilizers such as MgO and Y_2O_3 were especially susceptible to chemical attack by acidic vanadium pentoxide. Scandia (Sc_2O_3) and india (In_2O_3) in particular were examined in detail (Jones et. al.[5] Sheu et. al.[6]). Of these, india was found to be the most resistant to chemical attack by vanadium pentoxide.

<u>India stabilized Zirconia as a TBC coating:</u> Although india stabilized zirconia shows promise due to its relative inertness in vanadia containing atmospheres, there are still significant drawbacks in its use as a TBC material. First, india volatilizes at a lower temperature than zirconia. This resultin significant challenges for applyingplasma sprayed TBC's[1]. Although india stabilized zirconia coatings have been made in the t' phase (Sheu[6]), concerns about the volatility of indium oxide raise questions about the ability of india stabilized zirconia to form a homogenous coatings.

In_2O sublimes at 600 °C	10^{-4} torr at 650°C
In_2O_3 sublimes at 850 °C	10^{-4} torr at 850 °C

Jones, Reidy and Mess[5] were able to co-stabilize zirconia with yttrium oxide and indium oxide using a sol gel process. However, no attempt was made to provide ingot feedstock of this composition for EB-PVD testing. Furthermore, the high cost (> $300/kg) of In_2O_3 has also been a barrier for further research and development efforts.

Therefore, a logical approach was to incorporate the indium oxide into the ingot in a form that would make the indium oxide less volatile, therefore minimizing incidents of spitting, pressure fluctuations, and increase coating homogeneity while still providing enhanced corrosion resistant coating solely consisting of the t' phase. The strategy was to pre-react the indium oxide with a lanthanide oxide which forms either the $LnInO_3$ perovskite (La, Nd or Sm) or the hexagonal $LnInO_3$ (Gd or Dy). If the ingot contains zirconia and the $LnInO_3$ or just partially stabilized zirconia without free indium oxide, it was believed that a more homogeneous corrosion resistant coating could be deposited by electron beam physical vapor deposition (EB-PVD).

Advantages of Indate pre-cursor:

1) Perovskite indates ($LnInO_3$) are refractory compounds. The electropositive lanthanide ion (also stabilizers of the t' phase) stabilizes the In^{3+} state. It is the reduction to In^{1+} that leads to the volatilization of In.
2) Multiple stabilizing ions reduce thermal conductivity. The work of R. Miller [7] showed that TBC thermal conductivity decreases when numerous ions of different ionic sizes, valence and ionic weights are simultaneously incorporated into the zirconia as stabilizing agents. These are often referred to as oxide dopant clusters.

Lanthanide Selection: There are numerous factors that will determine the selection of the lanthanide ion accompanying the indium oxide.

1) Range of metastable t' phase field. Ideally one would like the largest rangepossible. Sasaki[8] found the t' phase between 15 and 20-mol % In_2O_3 when quenched from temperatures above 1500°C. Ideally this phase region would accompany the In mol% alone as well as the entire range up to the (Ln + In) mole percentage.
2) Melting temperature of $LnInO_3$ compound. The more refractory the compound, the better is the performance
3) Acidity/basicity of lanthanide ion. If La is used, this is likely to be strongly attacked by vanadium because of its basicity. As we progress through the heavier lanthanides (left to right on periodic table), the basicity decreases.
4) Ionic size and weight. Y is of the ideal atomic size for decreasing the monoclinic-tetragonal transformation temperature in ZrO2. (Sasaki [8]1993). As we move to smaller ions or larger ions this change in the transformation temperature is decreased. In addition, the greater the difference in ionic size and ionic weight between the In^{3+} and the Ln^{3+} ions, the lower the thermal conductivity (Miller[7]2004).

Phase Diagram Information: Only one ternary phase diagram exists containing any Ln_2O_3-In_2O_3-ZrO_2 ternary systems. That one is for Ln=Pr and it was produced by Bates [9]et.al in 1989. The compatibility relationships expressed in this diagram suggest that $PrInO_3$ perovskite would react with zirconia to form the $Pr_2Zr_2O_7$ pyrochlore and free indium oxide, the exact situation one should avoid. In addition, it has been shown[10] that the larger lanthanide ions (La-Gd) in zirconate pyrochlores react with the thermally grown oxide to form undersirable lanthanide aluminate phases. Therefore, the authors investigatedLn ions that formed stable binary oxides of the perovskite structure with In_2O_3 but did not form the pyrochlore structure or formed the pyrochlore structure sluggishly. Like the formation of the indate perovskites, the stability of the pyrochlore phase decreases as we proceed from the light to heavy lanthanides. The lanthanides of greatest interest are therefore Sm, Gd and Dy.

Sm_2O_3	Forms $Sm_2Zr_2O_7$ pyrochlore Stable to 1800°C (Yokakawa [11]1992)	Forms $SmInO_3$ perovskite (Schneider, Roth and Waring [12]1961)
Gd_2O_3	Forms $Gd_2Zr_2O_7$ pyrochlore Stable to 1575°C (Yokakawa [11]1992)	Forms hexagonal $GdInO_3$ (Schneider, Roth and Waring [12]1961)
Dy_2O_3	Does not form $Dy_2Zr_2O_7$ pyrochlore (Pascual and Duran[13] 1980)	Forms hexagonal $DyInO_3$ Stable to 1600 C (Schneider, Roth and Waring [12]1961)

Lanthnides heavier than Dy do not form either the pyrochlore[13] or binary indate phases[12]. The samarium series is of interest because the indiate perovskite forms and since Sm is the most electropositive ion of the lanthanide series (to prevent In^{1+} formation and volatilization); however, Sm also forms the most stable pyrochlore which is undesirable. Conversely, the dysprosium series is of interest because it does not form the pyrochlore zirconate or the perovskite structure. The hexagonal compound that does form is unstable above 1600°C. Therefore the challenge is to find a compound indium oxide precursor that will prevent indium volatilization but will not react with zirconia to form a pyrochlore and thus liberate free (and volatile) In_2O_3.

In 2007, Mohan et. al.[14] reported that in addition to forming the zircon YVO_4 phase that YSZ will react with vanadate salts below 747°C to form the zirconium pyrovanadate (ZrV_2O_7) phase. The role this phase plays in the mechanical properties of YSZ coatings containing vanadium warrants further study.

EXPERIMENTAL

$LnInO_3$ materials were synthesized by blending yttrium, samarium, gadolinium or dysprosium oxides (loss on ignition determined at 1300°C for all starting oxides) with indium oxide in a ball-mill with yttria-stabilized zirconia (YSZ) media at 55% solids loading without dispersants for 4h. The slurry was pan dried and calcined at 1300°C for 8h. X-ray diffraction was used to evaluate the phase purity of the material by comparing with the appropriate JCPDS cards. If the reaction was incomplete, the milling and calcinations were repeated. The fully-reacted lanthanide indate compositions were then ball-milled with YSZ media until the median particle size was 2 microns or less.

Table I. - Physical and Chemical Properties of the Fired Ingot Material

Ingot Material	Fired Density	Phase Content	Evaporation Quality
6 mole% $SmInO_3$	4.81 g/cc	t-ZrO_2, m-ZrO_2 + $LnInO_3$	Poor - Spitting
6 mole% $GdInO_3$	4.85 g/cc	t-ZrO_2, m-ZrO_2 + $LnInO_3$	Poor - Spitting
6 mole% $DyInO_3$	4.80 g/cc	t-ZrO_2, m-ZrO_2 + $LnInO_3$	Extremely Poor
6 mole% $SmInO_3$ +3 mole% Y_2O_3	4.59 g/cc	t-ZrO_2, m-ZrO_2 + $LnInO_3$	Poor –Spitting
6 mole% $GdInO_3$ +3 mole% Y_2O_3	4.63 g/cc	t-ZrO_2, m-ZrO_2 + $LnInO_3$	Poor – Spitting

The indate precursors were then blended with zirconia to the desired composition and formed by cold isostatic pressing into the EB-PVD ingots. The materials were heat treated between 1430 °C and 1530°C for 10h to achieve a theoretical density between 60 and 70%. Table I shows the fired densities, the phase content and the evapoaration quality of the ingot material as a function of the chemical composition. XRD revealed the fluorite structure along with residual monoclinic zirconia and the indate perovskites as listed in Table I.

The ingots were evaporated onto platinum aluminide coated MAR-M-247 nickel based alloy one inch diameter buttons in an industrial prototype EB-PVD coating system at Penn State University. XRD and SEM microstructures were prepared for each coating, with selectEDX presented for semi-quantitative coating chemistry analysis.

Corrosion reactivity tests were performed by reacting the coated coupons with a thin coating of vanadium pentoxide and heated to temperatures between 400 – 650°C for 4 – 6 hours. X-ray diffraction was performed on the pre-reacted and as-reacted coating to identify any phases forming due to the reaction with vanadium pentoxide.

RESULTS

1) Evaporation: In general, the ingots evaporated poorly in the industrial scale EB-PVD coating unit. The material showed "spitting" and extensive cracking during evaporation. The spitting is most likely due to the difference in the vapor pressure between zirconium oxide and indium oxide containing phases in the ingot, but can also be the result of localized differences in ingot densities and degree of connected porosity. Cracking can also occur if the ingot density is too high or the ingot does not have sufficient thermal shock resistance. Despite the difficulties during ingot evaporation, coatings were obtained for each material studied. However, it should be noted that some "spits" or coating defects were observed on the surface of the coated coupons. Lastly, yttrium oxide was added into the composition as an evaporation aid during powder formulation and ingot fabrication, but it did not appear to substantially improve ingot evaporability.

2) Coating Properties: XRD revealed that all of the coatings were single phase with the desired t' structure. The coating microstructure as observed by scanning electron microscopy revealed a columnar microstructure typical of those applied by the EB-PVD process. Figure 1 shows an SEM

micrograph of the 6 mol% GdInO3 stabilized zirconia coating surface morphology. In addition, EDX was performed on the coating surface to determine semi quantitative compositional information regarding traces of rare earth and indium oxide compositions. These results are listed in Table II.

The first measure of success was to obtain a coating which contained the acidic stabilizer In_2O_3. Table II compares the ease of evaporability and the relative amount of india within the coating for the various compositions studied. The two compositions containing samarium indate showed the highest amounts of residual indium followed by the sample containing both gadolinium and indium oxide. The ingot starting with 6 mole % indium oxide showed moderate amounts of indium remaining in the EDX trace although considerably less than either samarium containing composition despite starting with double the amount of indium oxide in the ingot.

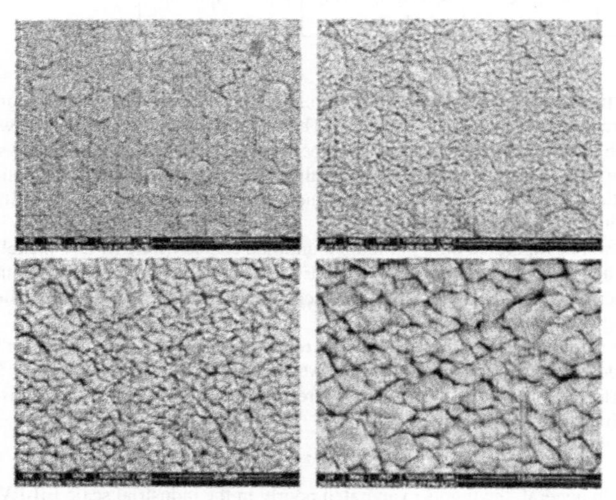

ESEM images showing the surface morphology of ZrO_2/Y/$GdInO_3$ deposited on a platinum aluminide bond coated MAR-M-247 button. Sample # S050923-1H
10/4/2005

Figure 1: SEM image of surface morphology of the EB-PVD coating obtained by evaporation of the 6 mol% $GdInO_3$-3 mol% Y_2O_3doped zirconia ingot composition. The coatings were applied on a platinum aluminide coated nickel base alloy. The top images show a lower magnification than the bottom images

3.) Reactivity Tests: Table III shows the results of the vanadium pentoxide reactivity tests. X-ray diffraction was performed on the various coatings before and after the reactivity tests in order to determine whether the coatings reactive with vanadium oxide. If any reactions occurred. the phases were identified. The sample containing samarium indate showed only the tetragonal prime phase until 500°C at 16h. Traces of the $LnVO_4$ phase with the zircon structure were observed in the samples containing the gadolinium and dysprosium indate at 400°C at 4 hours. Upon further testing at 500°C when exposed to vanadium oxide, traces of monoclinic zirconia and the ZrV_2O_7 phase appearedfor $DyInO_3$ containing samples. With the exception of the coating that contained 6mol% $GdInO_3$, the t' phase completely disappeared at 650°C suggesting that these coatings reacted with the vanadium oxide to destabilize the yttria stabilized zirconia. The 6mol% $GdInO_3$ composition showed the most promising results with regards to resistance against vanadium oxide attack.

Table II: A comparison of properties for ingots of various compositions studied. Ease of evaporation, EDX In_2O_3 content, SEM microstructure and the phase content.

Composition	Ease of EB-PVD Evaporation (TD: theoretical density.)	Amount of In in coating	Microstructure	Coating Phase
6 mol% In_2O_3	Poor (62 % TD ingot)	some	TBD	t'
6 mol% $SmInO_3$	Poor (60 and 70% TD ingots)	most	Columnar	t'
6 mol% $SmInO_3$ 3 mol% Y_2O_3	Poor (60% TD ingot)	most	TBD	t'
6 mol% $GdInO_3$	Poor (60 and 70% TD ingots)	little	Columnar – not homogenous	t'
6 mol% $GdInO_3$ 3 mol% Y_2O_3	Poor (60% TD ingot)	most	TBD	t'
6 mol% $DyInO_3$	Poor (70% TD ingot)	some	Poorly formed columns	t'

DISCUSSION

All of the india containing compositions were difficult to evaporate as an ingot. This makes it unlikely that these materials would be useful for EB-PVD applications. EB-PVD is typically used for aircraft engine coatings. This application would typically use clean fuels devoid of acidic corrosive impurities. The material may be more useful as a plasma sprayed powder, which is a more typical TBC form for the industrial gas turbine industry with increased probability of being exposed to vanadium.

The samarium containing compounds showed the most residual india in the coating as determined by EDX. Sm is the most electropositive of the lanthanide co-stabilizers and is less likely to form pyrochlores than the lighter lanthanides like La or Nd. The Sm and Gd containing materials formed the typical columnar microstructure while the Dy containing sample showed poorly formed columns. It is not clear whether this was the result of processing difficulties caused by the ingot composition or phase stability.

The reactivity test showed that of all of the lanthanide co-stabilizers, the $SmInO_3$ containing composition showed the highest onset temperature before $LnInO_3$ formation. This is at least partly a result of the higher india content in the evaporated coating. Along with the formation of the $LnInO_3$ and the expected monoclinic zirconia, the ZrV_2O_7 phase appeared as well. There is no evidence of any influence of the stabilizing agent on the formation of this phase. It is uncertain whether this phase has a role on the mechanical durability or lifetime of the TBC. In the Gd and Dy containing coatings (which showed lessincorporation by EDX) the onset temperature for the appearance of $LnInO_3$ was the same as that for the formation of the ZrV_2O_7 phase. The appearance of the zircon structure vanadate either at lower temperatures than or concurrently with the monoclinic zirconia suggests that mineralization reactions are not taking place.

Table III. Table listing the reaction temperatures and phases observed when exposed to vanadium oxide at elevated temperatures.

Composition	Reaction with vanadia at 400C/6h	Reaction with vanadia at 400C/16h	Reaction with vanadia at 500 C/8h	Reaction with vanadia at 500 C/16h	Reaction with vanadia at 600 C/8h	Reaction with vanadia at 600 C/16h	Reaction with vanadia at 650 C/16h
6 mol% SmInO3	t'	t'	t'	t' + SmVO4 + trace ZrV2O7	t' + SmVO4 + mono (ZrO2) + ZrV2O7	t' + SmVO4 + mono (ZrO2) + ZrV2O7	SmVO4 + mono (ZrO2) + ZrV2O7
6 mol% GdInO3	t'	t' + trace GdVO4	t' + trace GdVO4	t' + trace GdVO4	t' + GdVO4 + mono (ZrO2) + ZrV2O7	t' + GdVO4 + mono (ZrO2) + ZrV2O7	t' + GdVO4 + mono (ZrO2) + ZrV2O7
6 mol% DyInO3	t' + trace DyVO4	t' + trace DyVO4	t' + trace DyVO4	t' + DyVO4 + trace mono (ZrO2)	t' + DyVO4 + mono (ZrO2) + ZrV2O7	t' + DyVO4 + mono (ZrO2) + ZrV2O7	DyVO4 + mono (ZrO2) + ZrV2O7

CONCLUSIONS

It has been shown that indium oxide is an alternative stabilizing agent to yttria which does not react readily with vanadium oxide. A processing technique has been developed to incorporate increased amounts of indium oxide by using rare earth oxides by pre-reacting the indium oxide with samarium oxide or gadolinium oxide to form a stable perovskite ($GdInO_3$ or $SmInO_3$). This resulted in

reduced volatilization of the indium oxide and thus increased volume fractions of indium oxide being incorporated into the coating. Comparison of EDX data from evaporated coatings to the coatings produced after electron beam evaporation containing solely indium oxide and those containing $GdInO_3$ showed increased indium content present in greater quantities for those coatings containing the additional stabilizer. The primary findings of the presented work are summarized below:
1) That the addition of a lanthanide co-stabilizer (i.e.,Sm) will assist india incorporation into a EB-PVD thermal barrier coating. EDX revealed a greater india concentration in the 3 mol% coating as $SmInO_3$ than with 6 mol% In_2O_3.
2) The indate materials investigated in this effort do not appear to be ideal for EB-PVD coatings. This material combination is more likely to be better suited for plasma spraying.
3) Samples containing samarium indate showed the most resistance to reaction with vanadium pentoxide
4) The appearance of the $LnVO_4$ phase at temperatures below or concurrently with the monoclinic zirconia contra-indicates a mineralization reaction.

Continued efforts are suggested to further optimize the $LnInO_3$ content, to explore hot corrosion tests mimicking service conditions and to understand the role of the ZrV_2O_7 phase. In addition, additional efforts to prepare and field test plasma sprayed coatings of the india co-stabilized zirconia will be investigated. The materials described within are subject to a pending US patent.

REFERENCES

1) R.L. Jones **J Thermal Spray Technology** 6 [1] 1997 pp77-84
2) Hamilton and Nagelberg **JACerS** 67 [10] 1984 pp 686-690
3) Susnitzky, Hertl and Carter **JACerS** 71 [11] 1988 pp 992-1004
4) Jones and Mess **JACerS** 75 [7] 1992 pp1818-21
5) Jones, Reidy and Mess **JACerS** 76 [10] 1993 pp2660-2662
6) Sheu, Xu and Tien **JACerS** 76 [8] 1993 pp2027-2032
7) Miller, **Int. J. of Applied Ceramic Tech** 1 [2] 2004
8) Sasaki, Bohac and Gaukler **JACerS** 76[3] 1993 pp 689-698
9) Bates, Weber and Gatkin **Proc. – Electrochem. Soc. [Proc. Int. Symp. Solid Oxide Fuel Cells, 1st]** 1989 pp 141-156
10) C.G. Levi **Current Opinion in Solid State and Materials Science** 8 2004 pp77-91
11) Yokakawa, Sakai, Kaweda and Dokiya **Sci. Technol. Zirconia V [Int. Conf. 5th]** 1993 pp 59-68
12) Schneider, Roth and Waring **J. Res. Nat. Bur. Stds. Section A** 65 [4] 1961 pp345-374
13) Pascual and Duran **J. Mater.Sci.** 15 [7] 1980 pp1701-1708
14) Mohan, Yuan, Patterson, Desai and Sohn **JACerS** In Press

DAMAGE PREDICTION OF THERMAL BARRIER COATING BY GROWTH OF TGO LAYER

Y. Ohtake
IHI Corporation
1, Shin-Nakahara-Cho, Isogo-ku,
Yokohama-shi, Kanagawa 235-8501, Japan

ABSTRACT

A typical coating system is composed of thermal barrier coating (top coating) and environment barrier coating (bond coating), and it is applied to hot parts. The fracture mechanism of the coating system had been examined in two kinds of testing, heating testing in furnace at constant temperature and burner rig testing by cyclic thermal loading, in previous paper. The coating system fractured by spallation of top coating at the interface on bond coating. The occurrence of spallation of top coating depended on thickness of thermal growth oxidation (TGO) layer at the interface. The spallation of top coating was caused by TGO layer. The specimen was used top coating of thickness 0.5mm in burner rig testing, but the growth behavior of TGO layer might change if thickness of top coating was thin. This paper investigates to examine the growth behavior of TGO layer for cyclic thermal loading by specimen that is applied top coating of thickness 0.3mm. The thickness of TGO layer was measured from the observation at the interface in the specimen after the testing. The thickness of TGO layer increased as the number of cyclic thermal loading. It was also found that the growth law of TGO layer didn't change in cyclic thermal loading if thickness of top coating was thin. Thus, the growth of TGO layer in cyclic thermal loading could predict by an equation that was proposed in previous paper. The equation was made in two terms of the thickness of TGO layer and heating time, where heating time in the equation adopted total holding times at maximum temperature in cyclic thermal loading.

INTRODUCTION

Thermal growth oxidation (TGO) layer grows at the interface between thermal barrier coating (top coating) and environment barrier coating (bond coating) when two coating layers is used for coating system. One of the fracture types of top coating is spallation of top coating near the interface [1]-[12]. The cause is due to the growth of TGO layer. We can detect the growth simple because the growth of TGO layer occurs at the interface in top coating. Thus, we are necessary to develop growth law of the thermal growth oxidation (TGO) layer at the interface between top coating and bond coating. Top coating is applied for hot parts of airplane engine or gas turbine engine. The engine part is given cyclic thermal loading in the operation. Ohtake al. had examined the damage of rectangular plate specimen with top coating by burner rig testing [1]-[5]. It was found that the fracture type of top coating was spallation from observation of specimen after testing. The testing time of cyclic thermal loading takes several mouths for one specimen only. Ohtake had developed new equipment that can test hour specimens at same time in burner rig by using small circular plate specimen with top coating of thickness 0.5mm. An equation had proposed to predict the growth of TGO layer for cyclic thermal loading. However, the equation may change that the growth of TGO layer influences on thickness of top coating in cyclic thermal loading. This paper examined the growth of TGO layer after cyclic thermal loading by using small circular specimen with top coating of thickness 0.3mm. The growth law of TGO layer investigated for the influence of thickness of top coating.

EXPERIMENTAL PROCEDURE

The specimen is composed of a typical coating system and base metal. The coating system consists of top coating over bond coating. Base metal is single crystal CMSX-2 substrate of nickel base superalloy. Bond coating applies CoNiCrAlY that is manufactured by low pressure plasma spray (LPPS). Top coating applies 8 wt. percent yttria stabilized zirconia (YSZ) that is manufactured by air plasma spray (APS). Figure 1 shows the appearance in burner rig testing. Testing equipment in Fig.1 can test four circular plate specimens at same time by burner rig. The size of the circular plate of base metal is diameter 20mm and thickness 3mm. Top coating on the plate is thickness 0.3mm and bond coating thickness 0.125mm. The thickness of top coating becomes thin about 0.2mm than one of top coating in previous experiments [5]. The surface of the specimen is heated by high temperature gas and the back surface is cooled by air. The thermal history of one cycle is total time 3min, 20s heating time, 60s holding time and 100s cooling time. The maximum temperatures on surface of top coating are about 1473K, and then the temperature of the cooling surface are 1173K. The thermal loading is repeated until maximum 3000 cycles by burner rig. Those specimens are cut by diamond saw after the testing. TGO layer at interface in all specimens is observed by scanning electron microscope (SEM). The thickness of TGO layer is measured at ten points for one specimen and the average value is adopted for thickness of TGO layer of the specimen. Heating time is defined as the sum of

Fig.1 Appearance of thermal cycle test

holding time at maximum temperature in burner rig testing. The growth behaviors of TGO layer is examined from the relationship between thickness of TGO layer and heating time after cyclic thermal loading by burner rig testing.

EXPERIMENTAL RESULTS

Figure 2 shows the relationship between heating time and thickness of TGO layer, where heating time is total time at maximum temperature and the thickness of TGO layer is measured in the specimen at heating time in burner rig testing. TGO layer grows as the increase of heating time as shown in Fig.2 and the thickness increases as the number of cyclic thermal loading. The grown behavior doesn't change for the testing at constant temperature in furnace. Thus, it was found that the grown of TGO layer didn't depend on testing method and heating time was important factor to predict the grown of TGO layer. Equation (1) was proposed from the results of heating testing in furnace in previous paper [5]. The equation is expressed in terms of thickness of TGO layer w, heating time t and two constants k and n.

$$w = kt^n \qquad (1)$$

The line in Fig.2 is k=0.03 and n=0.3 in Eq.(1). The data of burner rig testing could predict by using the line of Eq.(1) in Fig.2. It was found that the growth of TGO layer in cyclic thermal loading could predict by Eq.(1).

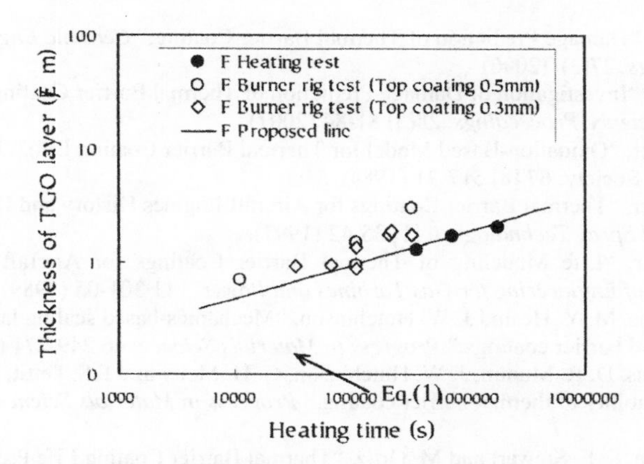

Fig. 2 Relationship between heating time and thickness of TGO layer

Crack was detected in top coat of the specimen after burner rig testing when the number of cycle reached more over 1000 cycles. The cyclic number of the occurrence of crack was didn't depend for thickness of top coating. The cracks existed in parallel direction of top coating surface from pore near the interface. It was considered that the delamination of top coating was occurred in the progress and the combination with a lot of cracks in top coating near the interface when TGO layer was increased with heating time.

CONCLUSIONS

This paper investigated the effect of the growth behavior of TGO layer for thickness of top coating in cyclic thermal loading. The burner rig testing was used to examine the behavior of TGO layer in cyclic thermal loading. The thickness of TGO layer after the testing increased as the number of cyclic thermal loading. The growth of TGO layer also denoted same tendency for the results of heating testing in furnace. It was found that the growth of thickness of TGO layer could predict by proposed equation in previous paper even if thickness of top coating was different. The equation was made in two terms of the thickness of TGO layer and heating time, where heating time in the equation adopted total holding times at maximum temperature in cyclic thermal loading.

REFERENCES

[1]Y. Ohtake, N. Nakamura, N. Suzumura and T. Natsumura, "Evaluation for Thermal Cycle Damage of Thermal Barrier Coating," *Ceramic Engineering and Science Proceedings*, 24(3) 561-566 (2003).

[2]Y. Ohtake, T. Natsumura, "Investigation of Thermal Fatigue Life of Thermal Barrier Coating," *Ceramic Engineering and Science Proceedings*, 25(4) 357-362 (2004).

[3]Y. Ohtake, T. Natsumura, K.Miyazawa, "Investigation of Thermal Fatigue Life Prediction of Thermal Barrier Coating," *Ceramic Engineering and Science Proceedings*, 26(3) 89-93 (2005).

[4]Y. Ohtake, "Damage Prediction of Thermal Barrier Coating," *Ceramic Engineering and Science Proceedings*, 27(3) (2006).

[5]Y. Ohtake, "Investigation of Damage Prediction of Thermal Barrier Coating," *Ceramic Engineering and Science Proceedings*, 28(3) 81-84 (2007).

[6]R. A. Miller, "Oxidation-Based Model for Thermal Barrier Coating Life," Journal of the American Ceramic Society, 67 [8] 517-21 (1984).

[7]R. A. Miller, "Thermal Barrier Coatings for Aircraft Engines History and Directions," *Journal of Thermal Spray Technology*, 6 [1] 35-42 (1997).

[8]. A. Miller, "Life Modeling of Thermal Barrier Coatings for Aircraft Gas Turbine Engines," *Journal of Engineering for Gas Turbines and Power*, 111 301-05 (1989).

[9]A. G. Evans, M. Y. He and J. W. Hutchinson, "Mechanics-based scaling laws for the durability of thermal barrier coatings," *Progress in Materials Science*, 46 249-271 (2001).

[10]A. G. Evans, D. R. Mumm, J. W. Hutchinson, G. H. Meier and F.S. Pettit, "Mechanisms controlling the durability of thermal barrier coating," *Progress in Materials Science*, 46 505-553 (2001).

[11]T. A. Cruse, S. E. Stewart and M. Ortiz, "Thermal Barrier Coating Life Prediction Model Development," Journal of Engineering for Gas Turbines and Power, 110 610-616 (1988).

[12]S. M. Meier, D. M. Nissley, K. D. Sheffler and T. A. Cruse, "Thermal Barrier Coating Life Prediction Model Development," *Journal of Engineering for Gas Turbines and Power*, 114 258-263 (1992).

YOUNG'S MODULUS AND THERMAL CONDUCTIVITY OF NANOPOROUS YSZ COATINGS FABRICATED BY EB-PVD

Byung-Koog Jang, Yoshio Sakka
Nano Ceramics Center, National Institute for Materials Science (NIMS)
1-2-1 Sengen, Tsukuba, Ibaraki 305-0047, Japan

Hideaki Matsubara
Japan Fine Ceramics Center (JFCC)
2-4-1 Mutsuno, Atsuta-ku, Nagoya, 456-8587, Japan

ABSTRACT

ZrO_2-4mol% Y_2O_3 coatings onto zirconia substrate were deposited by EB-PVD. Influence of the rotation speed of substrate and porosity on thermal conductivity of coated samples was investigated. Thermal conductivity of the ZrO_2-4mol% Y_2O_3 coatings was estimated by laser flash method. The thermal conductivity of coating layers shows decreasing tendency with increasing the porosity according to rotation speed. A nanoindentation technique for determining Young's moduli is applied to EB-PVD ZrO_2-4mol%Y_2O_3 coatings. Young's modulus of the surface regions of the coatings is found to be higher than that of the side regions.

INTRODUCTION

Thermal barrier coatings (TBCs) have received a large attention because they increase the thermal efficiency of gas turbine engines by increasing the gas turbine inlet temperature and reducing the amount of cooling air required for the hot section components. Among the various coating processes for producing TBCs, electron beam physical vapor deposition (EB-PVD) is widely used because it has several advantages of other techniques, including high deposition rate, use of high melting point oxides and excellent thermal shock resistance behavior due to columnar microstructure of the final coating [1-3].

A typical oxide for TBCs is partially stabilized zirconia containing 4 mol% yttria (4YSZ); this ceramic oxide has low density, low thermal conductivity, high melting point and good thermal shock resistance, i.e., excellent erosion resistance properties[4-6]. The investigation of thermal properties of coatings, especially thermal conductivity, is extremely important since the thermal efficiency of the coating, as judged by the temperature drop across the coatings, depends on the thickness of the coatings, its thermal conductivity, and the operating temperature.

In addition, characterization of the mechanical properties of the coatings is important for durability and developing superior thermal barrier coatings. Young's modulus is one of the most important properties for TBCs.

The purpose of this work is to investigate the influence of the rotation speed of substrate as well as porosity on thermal conductivity of EB-PVD ZrO_2-4mol% Y_2O_3 coatings by laser flash

method. In addition, this work describes Young's moduli of coatings using a Vickers nanoindentation technique, and to quantify the anisotropy in the vertical and horizontal directions of the coatings as a function of the substrate rotation speed during EB-PVD.

EXPERIMENTAL

EB-PVD coatings were obtained by deposition of 4 mol% Y_2O_3-stabilized ZrO_2 coatings onto disc-shaped zirconia substrates 10.0 mm in diameter and 1 mm thick. The coating process was carried out using commercial EB-PVD equipment (Von Ardenne Anlagentechnik). A coating chamber under 1 Pa vacuum was used with a commercially available 63 mm diameter YSZ ingot (Daiichi-kigenso Co.), and a 45 kW electron beam gun. Specimens were coated at different rotation speeds, namely stationary (0 rpm), 1, 5 and 20 rpm. The average coating thickness was about 300 μm and the substrate temperature was 950°C. The total porosity of the coatings was obtained by taking the difference between the density of the coated samples and the theoretical density of 4mol%Y_2O_3-ZrO_2.

Free standing coating layers were obtained by machining the substrate from the coated specimen prior to thermal conductivity measurements. The thermal diffusivity was determined using the laser flash method[7] using a thermal analyzer (Kyoto Densi, LFA-501). Prior to measuring the thermal diffusivity, the samples were sputter-coated with 1000 Å of silver to make them opaque, followed by a thin layer of colloidal graphite to ensure complete and uniform absorption of the laser pulse and similar surface radiative characteristics in all samples. All the measurements were carried out between room temperature and 1000°C at 200°C intervals in a vacuum chamber. Specific heat measurements of coated samples of 5mm diameter and 1.0 mm thickness were taken with a differential scanning calorimeter (Netzsch, DSC 404C) using sapphire as the reference material. The thermal conductivity of the coatings was then determined using

$$\lambda = \alpha \cdot C \cdot \rho \tag{1}$$

where λ is the thermal conductivity, α is the thermal diffusivity, C is the specific heat, and ρ is the density of the coatings.

In order to estimate the Young's moduli of the specimens experimentally, a nanoindentation method utilizing a Vickers diamond indenter was used at room temperature [8-10]. Each side of a sample was polished to yield a mirror face using diamond pastes. Measurements were performed at a constant load of 20 mN, loading time of 10 s and unloading time of 20 s by nanoindeatation equipment (Fischerscope, H100V). We measured Young's moduli in both the horizontal (surface region) and vertical (side region) directions of the coatings, because of the anisotropic columnar structure of the EB-PVD coatings. The so-called reduced modulus (E_r) was derived from the following equation, where A is indenter area, dp is the load increment and dh is the increment of the indentation depth in the range of 60~95% of maximum load for unloading after loading:

$$E_r = \frac{1}{2}\sqrt{\frac{\pi}{A}}\frac{dp}{dh} \tag{2}$$

Young's modulus of the coatings, E_c, can be extracted according to the following relation:

$$E_r = \left(\frac{1-\nu_i^2}{E_i} + \frac{1-\nu_c^2}{E_c}\right)^{-1} \tag{3}$$

where E_i and ν_i are the Young's modulus and the Poisson's ratio, respectively, of the indenter, and E_c and ν_c are the same quantities for the coatings. The microstructure of the coated samples was observed by SEM.

RESULTS AND DISCUSSION

Typical microstructures in the coatings observed from surface and side regions of EB-PVD YSZ coatings are shown in Fig. 1 for each substrate rotation speed. The morphology of surface regions of coatings shows the pyramidal morphology.

Figure 1. SEM micrographs of the surface regions and the side regions of ZrO_2-4 mol% Y_2O_3 coating specimens with different rpm of substrate during coating: (a) 0rpm, (b) 1 rpm, (c) 5rpm and (d) 20 rpm, respectively.

The 0 rpm layers exhibits sharp tetragon grains, whereas the rotated coatings reveal tetragon grains with the rounded edge. The morphology of surface regions of coatings deposited on substrates have a crystalline columnar texture with all columnar grains oriented in the same direction, namely perpendicular to the substrate, and with of a predominantly open microporosity. Some differences in the morphology of the columnar grains of specimens formed at different rotation speeds can be observed, however. The microstructure of the 0 rpm sample seems to be denser than that of specimens deposited at 1~20 rpm. In addition, well-developed columnar grains could be clearly observed, especially in the 20 rpm sample, whereas these were not noticeable in the 1 rpm sample. This is due, primarily, to the increase in the number of shadowed areas during deposition as the substrate rotation speed increases [11].

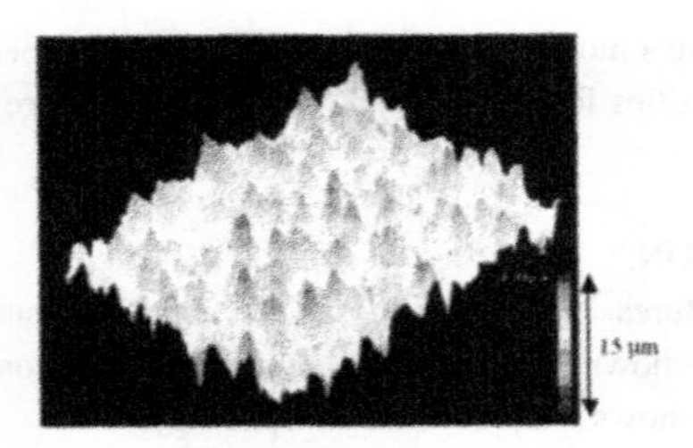

Figure 2. Laser microscope image of the top surface of coating layers. (Ra) is 2.61. μm.

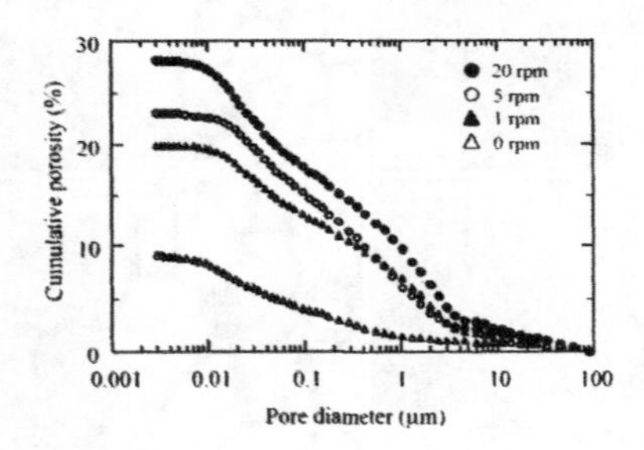

Figure 3. Porosity distribution for ZrO_2-4mol% Y_2O_3 coating layers by mercury porosimeter.

Fig. 2 shows an image of the top surface of the 5 rpm coating layers obtained by laser microscope. The deposited films have relatively rough surfaces because of the orientation of the crystal faces of the square pyramidal tips of the columnar grains. The columns of YSZ coatings are aligned in the same direction perpendicular to the substrate. It proves that the columnar grains have a uniform crystallographic texture in the growth direction. Fig. 3 shows pore distribution for the coating layers obtained using a mercury porosimeter. The pores range from nano-sized up to several microns. The volume of the nano sized pores of < 100 nm for 0 rpm sample is approximately 5 %. The number of nano-pores increases with an increase of substrate rotation speed, resulting in the increase of total the porosity. It seems that such nano pores exist mainly around feather-like features of the columns as well as intracolumnar pores of inside of the columns.

Fig. 4 shows the dependence of the thermal diffusivity on temperature up to 1000°C. At room temperature, the thermal diffusivity was found to decrease significantly with increasing substrate rotation speed primarily due to the increased porosity, which enhances phonon scattering. In addition, the thermal diffusivities of all coated samples decreased with increasing temperature.

A study of the influence of temperature on thermal diffusivity of sintered yttria-stabilized zirconia has been reported [12]. It was shown that the thermal diffusivity of sintered yttria-stabilized zirconia decreases with increasing temperature. Our results show that in addition to this temperature dependence, the thermal diffusivity also decreases with increasing porosity.

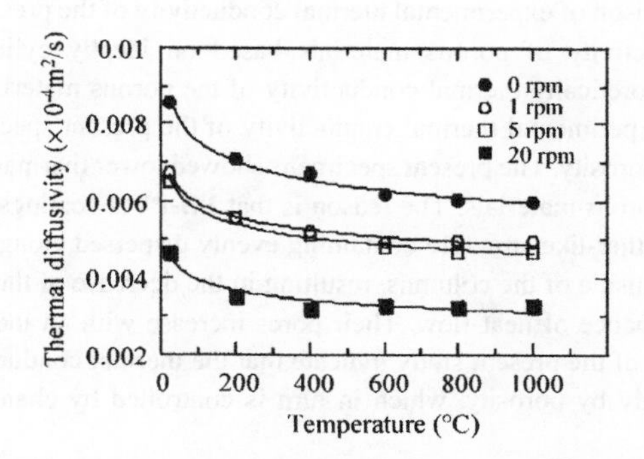

Figure 4. Thermal diffusivity vs temperature of ZrO_2-4mol%Y_2O_3 coating specimens obtained by EB-PVD.

Fig. 5 shows the results of thermal conductivity for coated specimens from room temperature to 1000°C. For coatings, thermal conductivity decreases slightly with increasing temperature. It is readily apparent that the thermal conductivities of the coatings are well below that of bulk, sintered material of composition 4 mol% Y_2O_3 stabilized ZrO_2, namely 2.59 W/m·k [7]. Plasma spray coating shows more low thermal conductivity than EB-PVD. This reason is related with lower porosity and lamellar pore structure in plasma coatings. At a given temperature, the thermal conductivity of the coatings decreases with increasing porosity as the rotation speed is increased.

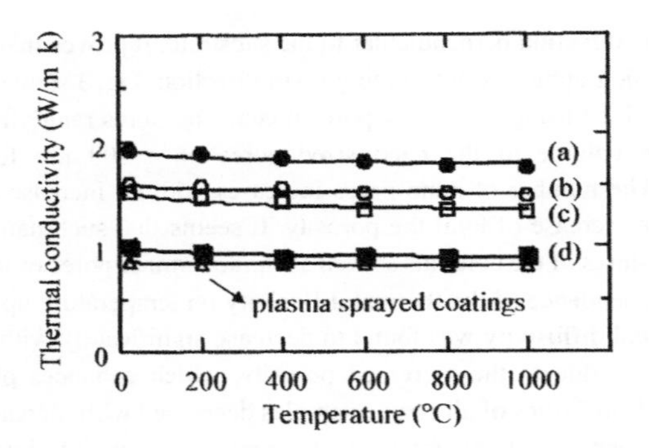

Figure 5. Thermal conductivity vs temperature of 4 mol% Y_2O_3-ZrO_2 coatings obtained by EB-PVD:(a) 0, (b) 1 , (c) 5 and (d) 20 rpm.

Fig. 6 shows the comparison of experimental thermal conductivity of the present specimens and theoretical thermal conductivity of porous materials based on ideally cylindrical pores and spherical pores [13]. The theoretically thermal conductivity of the porous materials decreases with increasing porosity. The experimental thermal conductivity of the present specimens remarkably decreases with increasing porosity. The present specimens showed lower thermal conductivity than theoretical values of the porous materials. The reason is that EB-PVD coatings consist of porous columnar grains with a feather-like structure containing evenly dispersed elongated pores as well as intracolumnar pores of inside of the columns, resulting in the decrease of thermal conductivity due to the effective disturbance of heat flow. Their pores increase with an increase of substrate rotation speed. The results of the present study indicate that the thermal conductivity of EB-PVD coatings is affected strongly by porosity, which in turn is controlled by changing the substrate rotation speed.

Based on this result, the porosity can be seen to be the dominant factor in obtaining a lower thermal conductivity for coatings. This is consistent with results showing that the porosity provides a major contribution to the reduction of the thermal conductivity of zirconia coatings [14, 15]. The thermal conductivity can usually be reduced by decreasing the mean free path due to phonon scattering at pores in the following manner: the thermal conductivity of a material describes the diffusivity of heat flow by phonon transport across a temperature gradient. The thermal conductivity (K) due to lattice vibration can be described by following expression[16]:

$$K = \frac{1}{3} C v \ell \qquad (4)$$

where C is the specific heat, v is the speed of sound and ℓ is the phonon mean-free path. From Eq. (4), if the phonon mean-free path is decreased, thermal conductivity will also decrease.

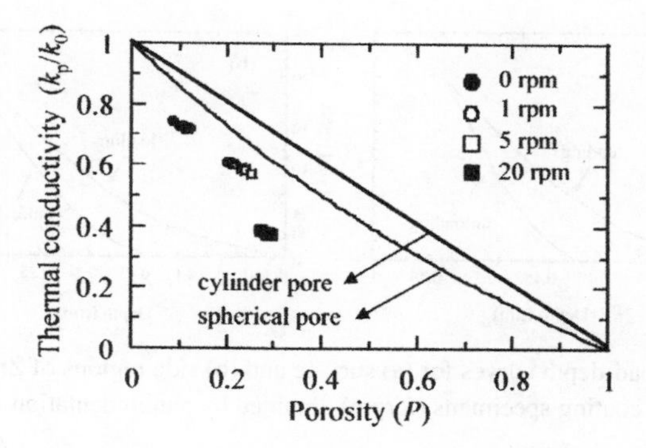

Figure 6. Thermal conductivity as a function of porosity for coatings specimens obtained by EB-PVD.

Fig. 7 presents typical load-depth curves obtained by nanoindentation for EB-PVD YSZ coatings specimen. The load is defined as the total force on the indenter and depth is the according to displacement is measured from the indenter's starting position. The load and the displacement into the surface are continuously measured during loading and unloading. Young's modulus can be calculated based on the unloading part of the curve as the unloading is a purely elastic recovery process. The slope of curve in the surface region of the coatings is larger than that of the side region. Different slopes of the load-depth curves indicate different Young's moduli.

Fig. 8 shows the influence of substrate rotation speed on the measured Young's moduli of both the surface and side regions of coatings. The Young's modulus of the side region was measured at a height of 50 μm from the substrate. The Young's modulus values show marked anisotropy in the coatings structure. The Young's modulus of the surface regions of the coatings is higher than that of the side regions. The Young's moduli of the surface regions have a nearly constant value of 230~250 GPa regardless of substrate rotation speed.

However, the Young's moduli of the side regions of the coatings show significant dependence on the substrate rotation speed. The Young's modulus of the side region of the 0 rpm coatings is higher than that of the rotated specimens. Furthermore, the Young's moduli of the sides of rotated specimens tend to decrease with an increase in rotation speed. This result means that the Young's modulus of each coating specimen strongly depends on the amount of porosity and the microstructure, which are both strongly dependent on the substrate rotation speed. The decrease in Young's modulus can be explained by the fact that the microstructures of rotated specimens become more columnar, with feather-like structures and an increased number of intracolumnar pores in the columns [17,18].

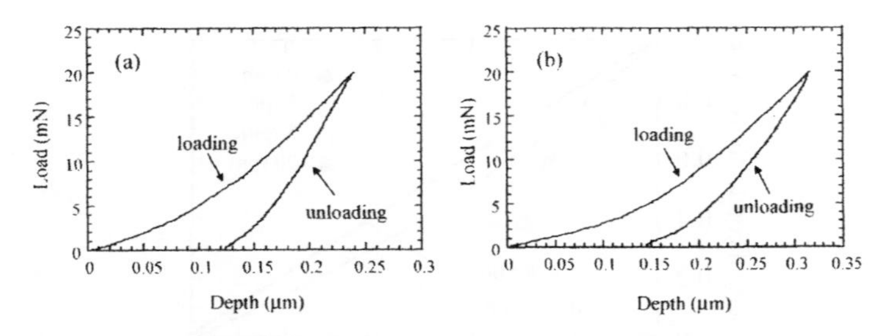

Figure 7. Load-depth curves for (a) surface and (b) side regions of ZrO_2-4 mol% Y_2O_3 coating specimens (5 rpm) obtained by nanoindentation.

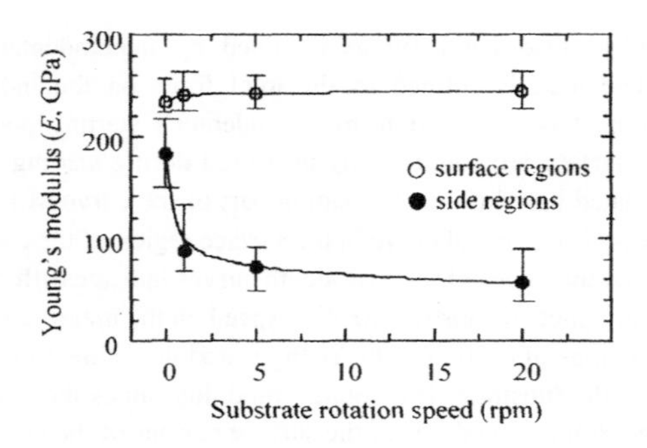

Figure 8. Young's modulus as a function of substrate rotation speed for ZrO_2-4 mol% Y_2O_3 coatings.

CONCLUSIONS

A ZrO_2- 4 mol% Y_2O_3 coating layer was deposited on ZrO_2 substrate by EB-PVD and found to consist of columnar grains. The porosity of coatings increased with increasing substrate rotation speed. The thermal conductivities of samples coated at different substrate rotation speed decrease with increasing temperature from room temperature to 1000°C. The porosity mainly leads to reduced thermal conductivity by the reduction of mean free path by phonon scattering.

Load-depth curves can be measured by nano indentation with a depth-sensing indenter and used to determine Young's moduli of EB-PVD coatings. Coatings were found to exhibit a strong anisotropy due to the textured microstructure. The Young's modulus of the surface regions of coatings is higher than that of its side regions. The Young's modulus of the side regions of coatings decreased with an increase in rotation speed.

ACKNOWLEDGMENTS

The authors acknowledge the financial support of the New Energy and Industrial Technology Development Organization (NEDO), Japan.

REFERENCES

[1] U. Schulz, B. Saruhan, K. Fritscher and C. Leyens, "Review on advanced EB-PVD ceramic topcoats for TBC applications," *Int. J. Appl. Ceram. Technol.*, **1**, 302-15 (2004).

[2]C. G. Levi, " Emerging materials and processes for thermal barrier systems," *Current Opinion in Solid State and Mat. Sci.*, **8**, 77-91 (2004).

[3]D. D. Hass, P. A. Parrish and H. N. G. Wadley, "Electron beam directed vapor deposition of thermal barrier coatings," *J. Vac. Sci. Technol.*, **16**, 3396-01 (1998).

[4]J. Singh, D. E. Wolfe, R. A. Miller, J. I. Eldridge and D. M. Zhu, "Tailored microstructure of zirconia and hafnia-based thermal barrier coatings with low thermal conductivity and high hemispherical reflectance by EB-PVD," *J. Mater. Sci.*, **39**, 1975-85(2004).

[5]O. Unal, T. E. Mitchell and A. H. Heuer, "Microstructure of Y_2O_3-stabilized ZrO_2 electron beam-physical vapor deposition coatings on Ni-based superallys," *J. Am. Ceram. Soc.*,**77**, 984-92 (1994).

[6]T. J. Lu, C. G. Levi, H. N. G. Wadley and A. G. Evans, "Distributed porosity as a control parameter for oxide thermal barriers made by physical vapor deposition," *J. Am. Ceram. Soc.*, **84**, 2937-2046 (2001).

[7]B. K. Jang, M. Yoshiya, N. Yamaguchi and H. Matsubara, "Evaluation of thermal conductivity of zirconia coating layers deposited by EB-PVD," *J. Mater. Sci.*, **39**, 1823-1825 (2004).

[8]E.lugscheider, K. Bobzin, S. Bärwulf, A. Etzkorn, "Mechanical properties of EB-PVD-thermal barrier coatings by nanoindentation," *Surf. Coat. Technol.*, **138**, 9-13 (2001).

[9]P. Mounaix, P. Delobelle, X.Melique, L.Bornier, D. Lippens, "Micromachining and mechanical properties of GaInAs/InP microcantilevers," *Mater. Sci. Eng. B*, **51**, 258-262 (1998).

[10]A.K.Bhattacharya, W.D. Nix, "Analysis of elastic and plastic deformation associated with indentation testing of thin films on substrates," *Int. J. Solid Struct.*, **24**, 1287-1298 (1988).

[11]B.K.Jang and H. Matsubara, "Microstructure of nanoporous yttria-stabilized zirconia films fabricated by EB-PVD," *J. Euro. Ceram. Soc.*, **26**, 1585-1590 (2006).

[12]D.P.H. Hasselman, L.F. Johnson, L.D. Bentsen, R. Syed, H.L. Lee, M.V. Swain, "Thermal diffusivity and conductivity of dense polycrystalline ZrO_2 eramics: A survey," *J. Am. Ceram. Bull.*, **66**, 799-806 (1987).

[13]H. Szelagowski, I. Arvanitidis and S. Seetharaman, "Effective Thermal Conductivity of Porous Strontium Oxide and Strontium Carbonate Samples," *J. Appl. Phys.*, **85**, 193-198 (1985).

[14]T.J. Lu, C.G. Levi, H.N.G. Wadley, A.G. Evans, "Distributed porosity as a control parameter

for oxide thermal barriers made by physical vapor deposition," *J. Am. Ceram. Soc.*,84, 2937-2946 (2001).
[15]J. Singh, D.E. Wolfe, "Review nano and macro-structured component fabrication by electron beam-physical vapor deposition (EB-PVD),"*J. Mater. Sci.*, **40**, 1-26 (2005).
[16]P.G.Klemens, "Theory of heat conduction in nonstoichiometric oxides and carbides," High Temp.-High Pressure, **17**, 41-45 (1985).
[17]U.Schulz, S.G. Terry, G. Levi, "Microstructure and texture of EB-PVD TBCs grown under different rotation modes,"*Mater. Sci. Eng. A*, **360**, 319-329 (2003).
[18]C.A. Johnson, J.A.Ruud, R.Bruce, D.Wortman, "Relationships between residual stress, microstructure and mechanical properties of electron beam–physical vapor deposition thermal barrier coatings," *Surf. Coat. Technol.*, **108-109**, 80-85 (1998).

INFLUENCE OF POROSITY ON THERMAL CONDUCTIVITY AND SINTERING IN SUSPENSION PLASMA SPRAYED THERMAL BARRIER COATINGS

H. Kaßner, A. Stuke, M. Rödig, R. Vaßen, D. Stöver,
Forschungszentrum Jülich, Institut für Energieforschung 1, IEF-1
Jülich, NRW, Germany

ABSTRACT

Suspension plasma spraying (SPS) was investigated as a potential manufacturing route for thermal barrier coatings. In this process powders with a particle size typically between a few up to several hundred nanometres are dispersed and stabilized in a suspension and then injected into the plasma. So, the SPS process makes it possible to directly feed nano scaled particles into the plasma plume, in contrast to the standard APS process, in which powders with a particle size above 10 µm have to be used. The direct processing of nano particles by the SPS process leads to new microstructures and properties. The size range of the porosity is shifted to lower values and also the porosity levels can be increased easily by the use of the SPS technology.

Free-standing coatings made of yttria partially stabilized zirconia (YSZ) with different porosity levels were produced and the pore size distribution was measured by mercury porosimetry. The thermal conductivity was measured for different porosity levels. Additionally, the effect of these different porosity levels on the sintering was investigated by dilatometric measurements during annealing at high temperatures. Furthermore the thermal expansion coefficient was determined from expansion during heating. For comparison, the results were compared to those of standard APS coatings.

Keywords: suspension plasma spraying, thermal barrier coating, yttria stabilized zirconia, thermal conductivity, thermal expansion coefficient, sintering

INTRODUCTION

The increase of the efficiency and performance of gas turbines is a great challenge in ecological as well as in economical perceptions. Both are related to the inlet temperature and the heat loss of the gas turbine.[1-3] For increasing typically thermal barrier coatings (TBC's) generated by atmospheric plasma spraying (APS) comprising a 0.25 to 0.5 mm thick yttria stabilized zirconia (YSZ) combined with an air cooling are widely used. The heat flux and hence the heat loss is mainly controlled by the thermal conductivity and the temperature gradient in the ceramic layer. The thermal conductivity mainly depends on the grain size, the phase composition, morphology and porosity.[4-8] Here the suspension plasma spraying (SPS) process offers new possibilities to adjust new microstructures with a low conductivity. By establishing the nanotechnology with its improved physical, mechanical and thermal properties also new ways for processing these small particles in the APS process were focused.[9-17] One possibility is their agglomeration to a particle size which enables a sufficient flowability. Nevertheless the semi-molten or molten droplets are still in the upper µm scale. Another possibility is suspension plasma spraying (SPS). Therefore a heterogeneous mixture of nano-particles together with a fluid is injected into the plasma plume. This enables the directly process of nano-particles. Thus molten droplets with a diameter of a few hundred nanometres to a few micrometers can be generated. Thereby also the generated splats are much smaller compared to conventional APS splats. Also the grain size of the ceramic is much lower. Hence new and often improved coating structures combined with a wide band of porosity levels could be generated. So the SPS process seems to be a promising technique for increasing the performance of conventional YSZ APS coatings.

In this work the influence of different porosity levels on the thermal conductivity, the thermal expansion coefficient and the sintering behaviour of different SPS coatings is investigated.

EXPERIMENTAL

For the experiments four different SPS coatings with different porosity levels and a thickness of about 300μm were produced. For comparison also an APS coating with a thickness of 300μm was sprayed. The coatings were produced with a Triplex II APS plasma gun supplied by Sulzer Metco AG, Wohlen, Switzerland. Square steel disks with a side length of 50 mm and a thickness of 2mm served as substrate. All substrates were sand blasted before coating. For the suspension preparation yttria stabilized zirconia (5YSZ) from Tosoh Corporation, Tokyo, Japan with a particle diameter of d_{50}=300 nm was used. The powder was dispersed in an ethanol based suspension. The mass content varied between 10 to 30 wt%. To achieve the required particle size all powders were dispersed in ethanol and ball-milled for at least 24h. As grinding stock zirconia or alumina balls with a diameter from 2 to 5 mm were used. The suspensions have been stabilized by the addition of 1.5 wt.% of a dispersant. For the APS standard coatings a 5YSZ from Sulzer Metco AG, Wohlen, Switzerland with a diameter of d_{50}=50μm was used. Particle size and distribution was measured with an Acoustic Spectrometer DT-1200 (Dispersion Technology Inc., Bedford Hills, USA) or an Analysette 22 (Fritsch GmbH, Idar Oberstein, Germany) using Fraunhofer diffraction. The microstructure was inspected in cross-sections or free standing coatings with a scanning electron microscopy (SEM), Ultra 55 (Carl Zeiss NTS AG, Germany). Dilatometric measurements were performed in a high-temperature dilatometer, Setsys 16/18, Setaram, Kep Technologies, France. The samples were cut from the freestanding layers. The sample length was 25 mm. A heating rate of 3 K/min up to 1200°C and dwell times at this temperature of 10 h was used. Pore size distributions in the layers were determined by Pascal 140 and 440 mercury porosimeters made by CE-Instruments, Milan, Italy, operating in a pressure range between 0.008 and 400 MPa, corresponding to pore diameters between 3.6 nm and 90 μm. Thermal diffusivity experiments were conducted using a laser flash device (Model: THETA, Netzsch, Germany) on disk-shaped specimens 12 mm in diameter and 1.5 mm in height. Measurements were performed at six different temperatures in the range 20–1200 °C, and repeated fives times at each temperature for statistical purposes.

RESLUTS AND DISCUSSION

Porosity

In sum four SPS and one APS coating with different porosity levels were generated. The pore size distribution of the SPS as well as the APS coating is typical bimodal. All SPS coatings have a higher number of pores all over the pore size distribution compared to the APS coating. Thereby the much higher porosity level of the SPS coatings are mainly induced due to the increased fine pore volume lower than 1 μm. Meanwhile fine pores radii with a size smaller than 1 μm are attributed to micro cracks and micro pores, the pore radii larger 1 μm are globular pores. The pore size distribution of all coatings can be seen in figure 1.

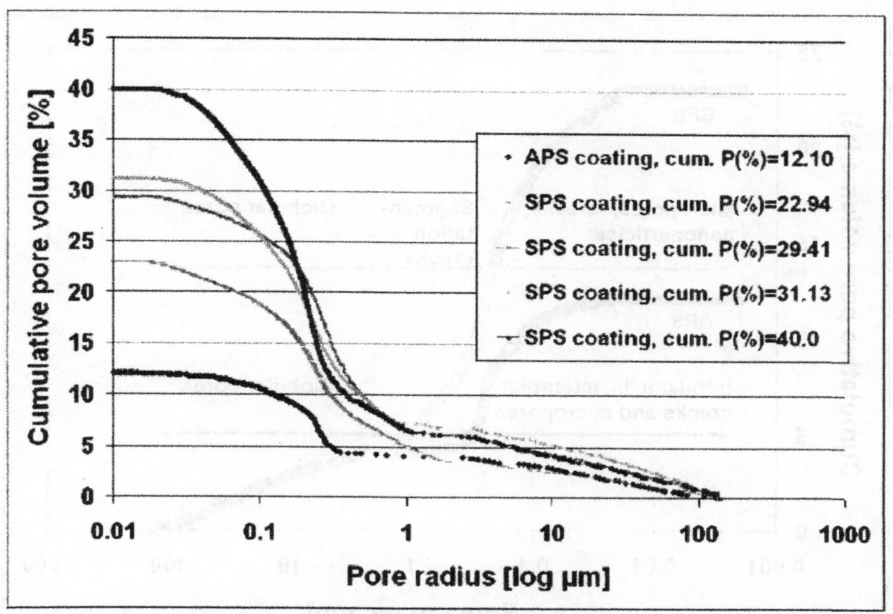

Figure 1. Porosity distribution of the SPS and APS coating as a result of Hg-porosimetry measurements.

The porosity level of the APS coating is at about 12.10%, the SPS coatings have porosity levels of up to 40.0%. The exact porosity levels of the different SPS coatings are 22.94%, 29.41%, 31.13% and 40.0%. The difference in the microstructure and the porosity levels of the SPS coatings can be attributed to different processing conditions like mass content of the suspension, plasma power and spraying. It has to draw out that there are different effects on the porosity levels for SPS and APS coatings. The pores lower 1 µm of the APS coating belong to intralamella and intersplat cracks. The pores of the SPS coatings are split into two areas. The pore radii between 0.2 µm to 1 µm are caused by segmentation cracks. Pores lower 0.2 µm mainly belong to micropores and embedded clusters of unmolten primary particles. Figure 2 points out the different curve progression of an APS and SPS coating.

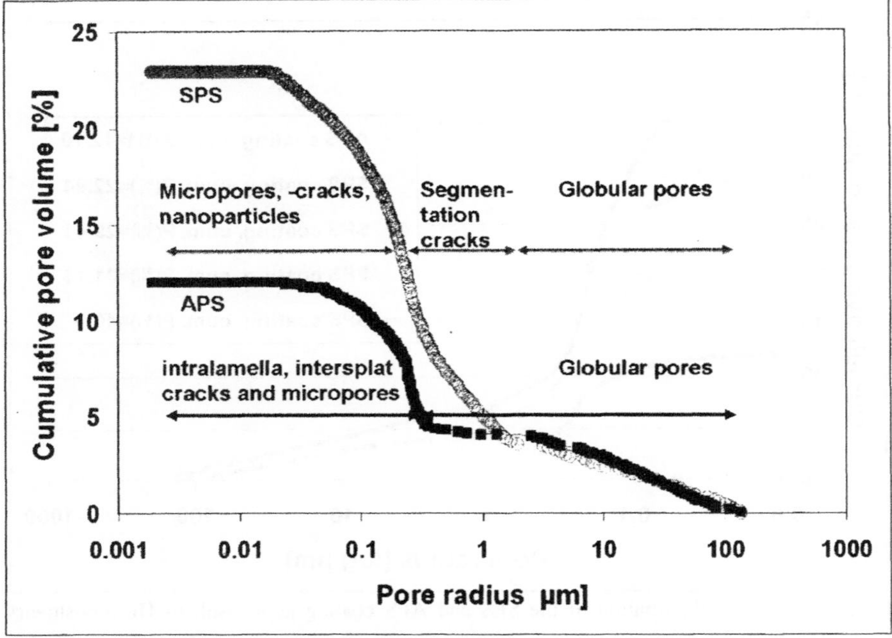

Figure 2: Curve progression of an APS and SPS coating.

Also the ratio between pores bigger than 1µm to pores lower than 1µm is different. The ratio of the APS coating is at about 0.5. For SPS coatings it is lower than 0.3. This means that SPS coatings contain a much higher amount of micro pores and micro cracks. Figure 3 depicts a cross section of the SPS coating with an open porosity of 22.94%. The visible small pores are isolated and homogenously distributed. Also clusters of unmolten nanoparticles can be seen. The big vertical crack belongs to a segmentation crack. Compared to the APS coating all SPS coatings show also a number of segmentation cracks.

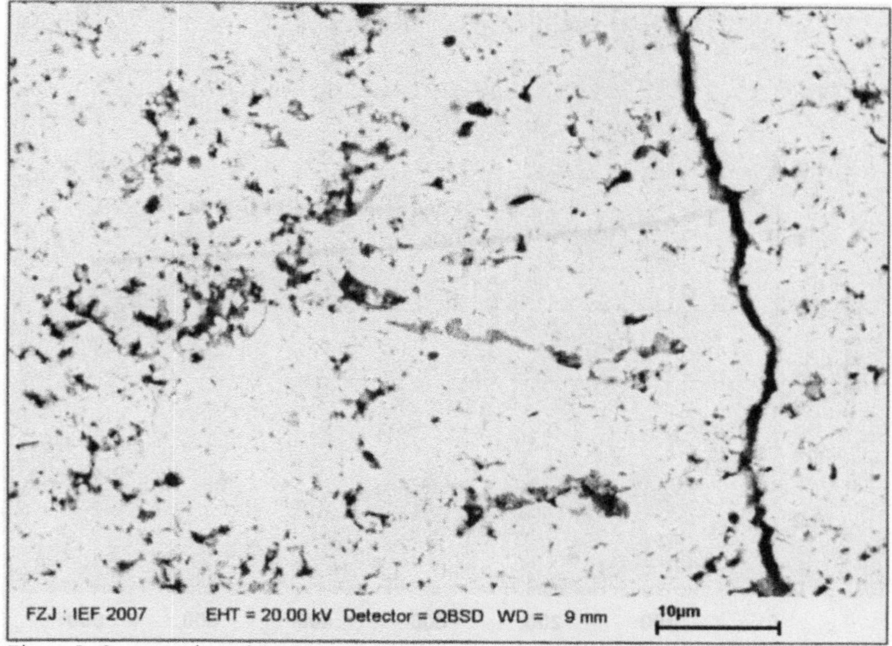

Figure 3: Cross section of the SPS coating with a porosity of 22.94%.

Sintering

Figure 4 illustrates the results of dilatometric measurements of the SPS sample with a porosity of 22.94%. After reaching the dewll temperature of 1200°C the sintering rate is rather high but decreases fast with time. The period at the end of the heating phase and at the beginning of the dwell time is shown in figure 5. The deviation from a straight line at the end of the heating at about 1100°C is a result of the onset of sintering. This effect is observed for all SPS and APS coatings.

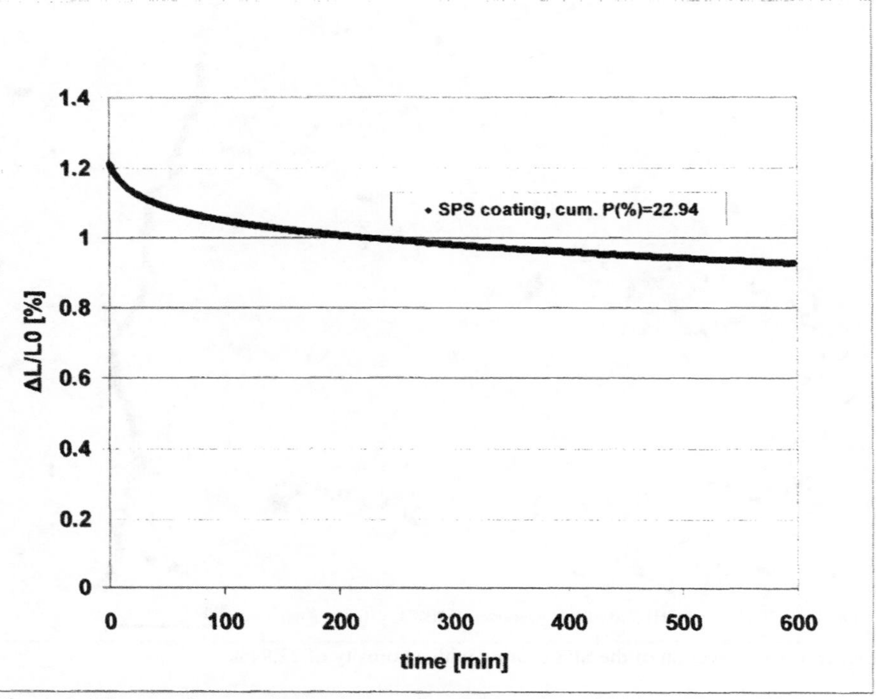

Figure 4: Dilatometric curve for the freestanding SPS coating with a porosity of 22.94%

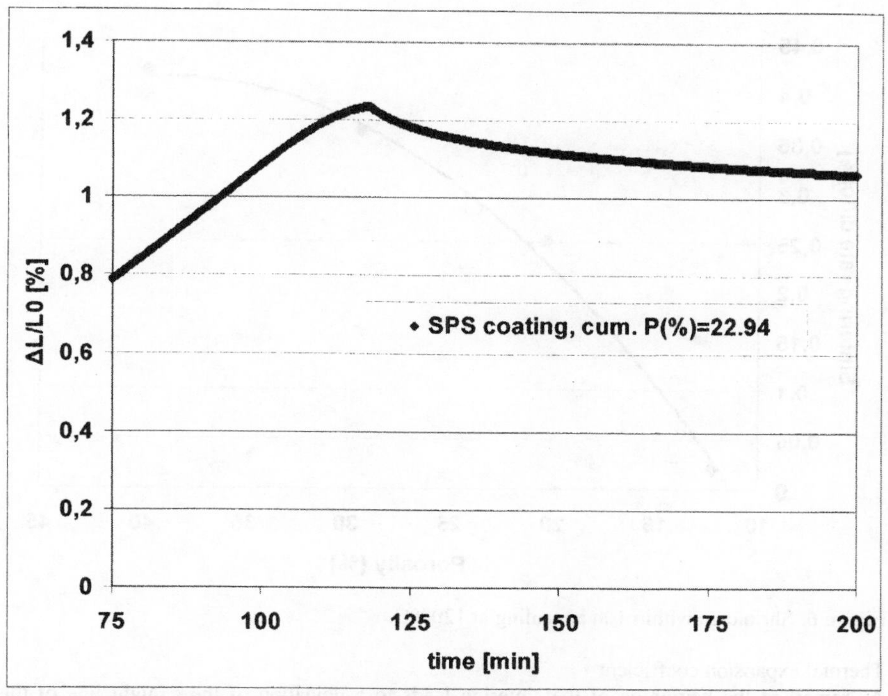

Figure 5: Period of the last minutes of heating phase and at the beginning of the dwell time

The effect of the porosity on the sintering of the SPS coatings after 10h is shown in figure 6. The sintering of the SPS coatings depends directly on the porosity level. Meanwhile the coating with a the lowest porosity of 22.94% features a sintering rate of 0.27% the coating containing a porosity of 40.0% has a sintering of 0.42% compared to the initial length. The APS coating shows a rather low sintering of 0.02%. Reasons for the higher sintering of the SPS coatings could be seen in the high amount of un-molten nanoscaled primary particles and the big number of micro- and nanopores in the SPS coatings. Both factors lead to an increased sintering especially at the beginning of the annealing. Another influencing parameter could be the higher silicon content in the nanopowder that serves as sintering aid.

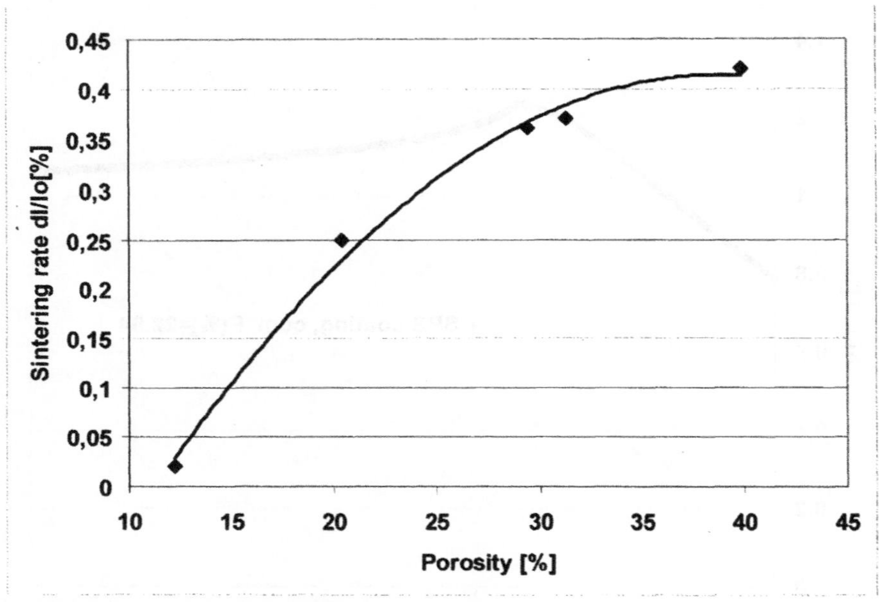

Figure 6: Shrinkage within 10h annealing at 1200°C.

Thermal expansion coefficient

As remarked the beginning of the sintering leads to a deviation of the straight line of the thermal expansion above 1100°C. The thermal expansion between room temperature and 1200 °C is almost linear for all coatings. The average value of thermal expansion coefficient (TEC) for the coating with 29.41% is determined to be 10.5×10^{-6} K^{-1}. Also the other porosity levels show a TEC between 10.5 to 10.9×10^{-6} K^{-1} independent from the porosity levels. The SPS coating with a density of 20.4% shows a thermal expansion coefficient of 10.6×10^{-6} K^{-1}. The TEC of the SPS coating with a porosity level of 31.3% is determined to be 10.9×10^{-6} K^{-1}. In contrast the APS coating reaches 11.2×10^{-6} K^{-1}. The thermal expansion data of the SPS coating with a porosity of 29.41% is shown in figure 7.

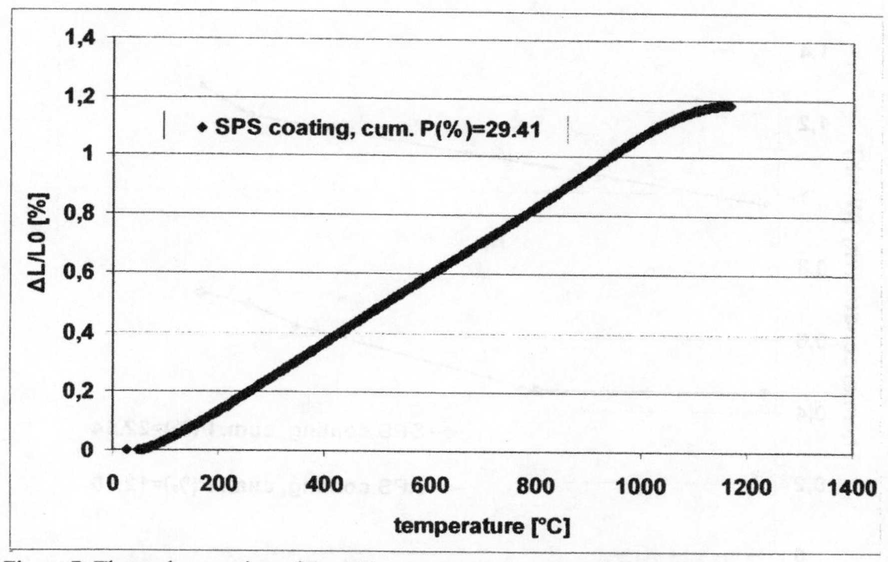

Figure 7: Thermal expansion of the SPS coating as a function of temperature

Thermal conductivity

The thermal conductivity was calculated with the equation $\lambda(T)=\alpha(T)\cdot Cp(T)\cdot \rho(T)$. Thereby the density $\rho(T)$ is the density of the porous material. The thermal diffusivity measurements of the SPS coatings reveal rather low values compared to the APS coating. The SPS coating with a porosity of 20.41% ranges from 0.0027 to 0.003 cm^2/s (from 25 to 1200°C) the APS coating has 0.0038 to 0.0035 cm^2/s (from 25 to 1200°C). Hence the calculated thermal conductivity of the SPS coatings is much lower as the standard APS coating. Figure 8 points out the influence of the temperature on the thermal conductivity for these two coatings as a function of the temperature.

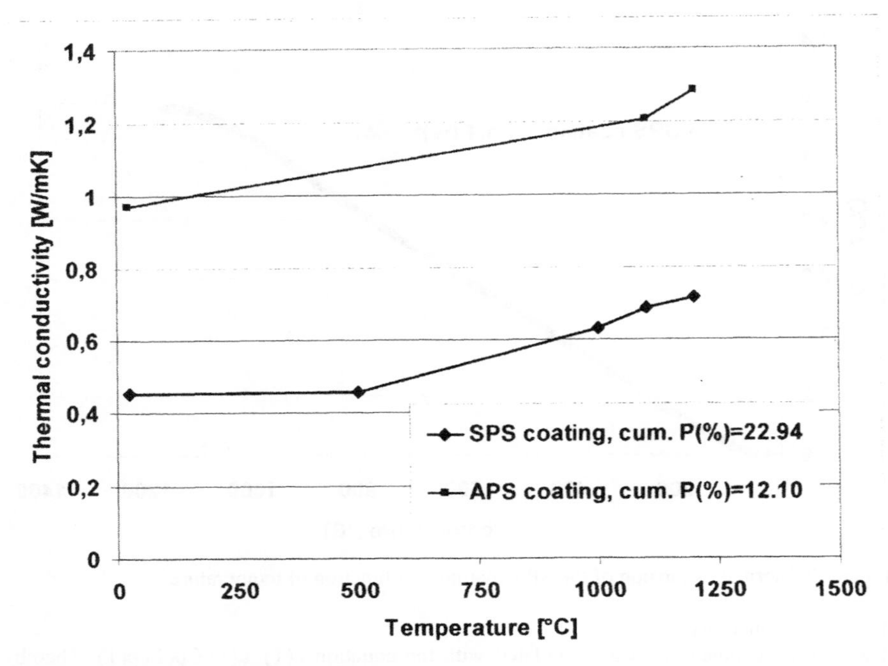

Figure 8: Thermal conductivity of an APS and SPS coating as a function of the temperature

The shape of the curves is for both coatings similar. Beginning from a low level at room temperature the thermal conductivity rises continuously with increasing the temperature. Nevertheless the values of the SPS coating are up to 50% less than that ones of the APS coating. The thermal conductivity at the technical relevant temperature of 1200°C for all coatings is given in table I.

Table I: Thermal conductivity at 1200°C

	porostiy level [%]	thermal conductivity [W/mK]
SPS	22.94	0.92
SPS	29.41	0.58
SPS	31.31	0.50
SPS	39.90	0.42
APS	12.12	1.28

There is a correlation between the porosity level and the thermal conductivity (figure 9). An increased porosity level leads to a decreased thermal conductivity. The thermal conductivity is 0.92 W/mK for the coating with a porosity of 22.94%. The coating with 39.9% only has a thermal conductivity of 0.42 W/mK. So a doubling of the porosity leads to a reduction of the thermal conductivity of 50%. The 1.3 W/mK at 1200°C of the APS coating is nearly 3 times higher.

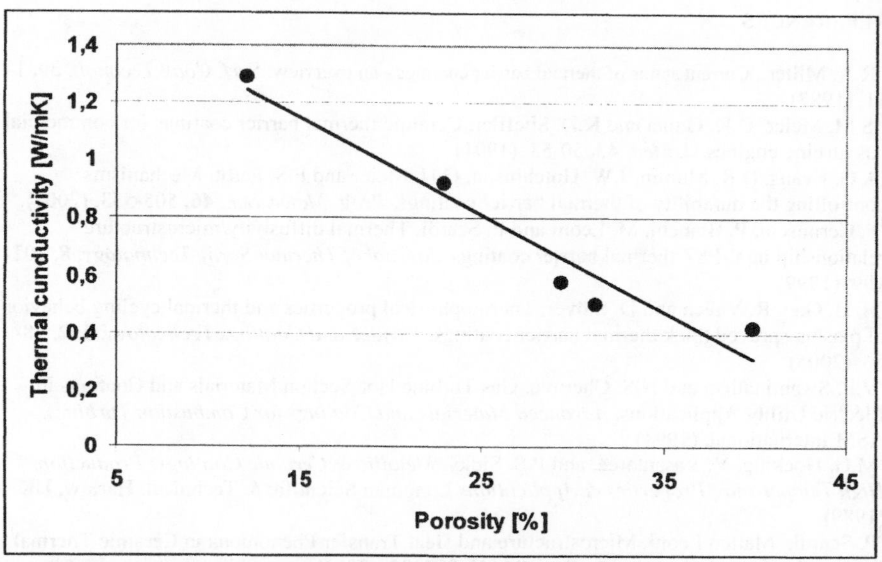

Figure 9: Influence of the porosity on the thermal conductivity.

CONCLUSIONS

As the results show, the SPS process allows improving the microstructure and the thermo physical properties of conventional coatings significantly.

Compared to the conventional APS process the SPS process enables a much wider band of porosity levels in thermal barrier coatings. Meanwhile conventional APS coatings normally have porosity levels of less than 15% the defined changing of the process parameters allows the SPS process to generate coatings with a porosity of up to 40%. Especially the amount of porosity with a pore radii lower than 1 μm induced by micro cracks, micro pores and some embedded un-molten nano particles is much higher compared to APS coatings.

The sintering curve characteristics are similar to conventional ones. The results also show that the sintering rate correlates directly with the porosity of the SPS coatings. So the coating with the lowest porosity level of 22.94% features a sintering rate of 0.27%, the coating a porosity level of 40.0% has a sintering rate of 0.42%. Reasons for the higher sintering of the SPS coatings could be seen in the high micro porosity and embedded unmolten primary particles in the coating. Also the higher silicon content in the nanopowder might enhance the sintering.

The thermal expansion and the thermal expansion coefficient are not influenced by the porosity levels. The thermal expansion coefficient varies between 10.6 to 10.9 $\times 10^{-6}$ K^{-1}. This is a little bit lower than that one of the APS coating with 11.2×10^{-6} K^{-1}.

As expected the high porosity levels in the SPS coatings lead to a decrease of the thermal conductivity by a factor of 3. Values of 0.42 W/mK can be obtained at the technical relevant temperature of 1200°C.

REFERENCES

[1] R.A. Miller., Current status of thermal barrier coatings - an overview, *Surf. Coat. Technol.*, **30**, 1-11, (1987).
[2] S.M. Meier, D.K. Gupta and K.D. Sheffler, Ceramic thermal barrier coatings for commercial gas turbine engines, *J. Me.t*, **43**, 50-53, (1991).
[3]A.G. Evans, D.R. Mumm, J.W. Hutchinson, G.H. Meier and F.S. Pettit, Mechanisms controlling the durability of thermal barrier coatings, *Prog. Mater. Sci.*, **46**, 505-553, (2001).
[4]F. Cernuschi, P. Bianchi, M. Leoni and P. Scardi, Thermal diffusivity/microstructure relationship in Y-PSZ thermal barrier coatings. *Journal of Thermal Spray Technology*, **8**, 102-109, (1999).
[5]H. B. Guo, R. Vaßen and D. Stöver, Thermophysical properties and thermal cycling behavior of plasma sprayed thick thermal barrier coatings, *Surface and Coatings Technology*, **192**, 48-56, (2005).
[6]V.P. Swaminathan and N.S. Cheruvu, Gas Turbine Hot-Section Materials and Coatings in Electric Utility Applications, *Advanced Materials and Coatings for Combustion Turbines*, ASM International, (1994).
[7]M.G. Hocking, V. Vasantaree, and P.S. Sidky, *Metallic & Ceramic Coatings: Production, High Temperature Properties & Applications*,Longman Scientific & Technical, Harlow, UK, (1989)
[8]P. Scardi, Matteo Leoni, Microstructure and Heat Transfer Phenomena in Ceramic Thermal Barrier Coatings, J. *Am. Ceram. Soc.*, **84** [4], 827–35, (2001).
[9]J.Oberste Berghaus, S. Bouaricha, J.-G. Legoux, C.Moreau, Injection conditions and in-flight states in suspension plasma spraying of alumina and zirconia nano-ceramics, *Thermal Spray 2005: Explore its surface potential!*, ASM International, 512 – 518, (2005)
[10]C. Delbos, J. Fazilleau, V.Rat, J.F. Coudert, P. Fauchais, B. Pateyron, Phenomena Involved in Suspension Plasma Spraying, Part 1: Suspension Injection and Behaviour, *Plasma Chem. And Plasma Processing*, **26**(4), 371-391, (2006)
[11]C. Delbos, J. Fazilleau, V.Rat, J.F. Coudert, P. Fauchais, B. Pateyron, Phenomena Involved in Suspension Plasma Spraying, Part 2: Zirconia particle treatment and coating formation, *Plasma Chem. And Plasma Processing*, **26**(4), 393-414, (2006)
[12]H. Gleiter, Nanostructured Materials: Basic concepts and microstructure, *Acta Materialia*, **48**, 1-29, (2000).
[13]D. Vollath, D.V.Szabó, Nanocoated Particles: A Special Type of Ceramic Powders, *Nanostructured Materials*, **4**(8), 927-938, (1994).
[14]R.S. Lima, A. Kucuk and C.C. Berndt , Evaluation of microhardness and elastic modulus of thermally sprayed nanostructured zirconia coatings. *Surf. Coat. Technol*,. **135**, 166–172 (2001).
[15]Y. Zeng, S.W. Lee, L. Gao and C.X. Ding , Atmospheric plasma sprayed coatings of nanostructured zirconia. *J. Eur. Ceram. Soc.*, **22**, 347–351, (2002).
[16]M. Gell , Application opportunities for nanostructured materials and coatings. *Mater. Sci. Eng.*, **A204**, 246–251, (1995).
[17]B.H. Kear and G. Skaudan , Thermal spray processing of nanoscale materials. *Nanostruct. Mater.*, **8**, 765–769(1997).

NUMERICAL INVESTIGATION OF IMPACT AND SOLIDIFICATION OF YSZ DROPLETS PLASMA-SPRAYED ONTO A SUBSTRATE: EFFECT OF THERMAL PROPERTIES AND ROUGHNESS

N. Ferguen[1], P. Fauchais[1], A. Vardelle[1] and D. Gobin[2]
[1] SPCTS UMR 6638, University of Limoges, 16 rue Atlantis, 87068 Limoges, France
[2] FAST – CNRS, Campus Universitaire, Bldg 502, 91405 Orsay, France

ABSTRACT

This study deals with numerical simulations of the impact and flattening of Yttria Stabilized Zirconia (YSZ) particles plasma sprayed onto smooth and rough substrates. On smooth substrates, predictions showed the effect of the thermal diffusivity of the substrate as well as the thermal contact resistance between splat and substrate on particle flattening and solidification. When solidification starts before flattening is completed, the contact between the flowing liquid and substrate in the periphery of the lamella becomes very poor and slows down the solidification process. On rough substrates, simulated by adding square columns on the smooth substrate surface, the distance between columns is a key parameter affecting the geometry of the lamella resulting from the droplet impact and, its solidification.

INTRODUCTION

The top-coat of the thermal barrier used for the protection of the hot sections components of gas turbine engines is conventionally deposited by either plasma spraying or electron beam physical vapor deposition (EBPVD). Although the search for new ceramics suitable for TBC continues, these techniques use Yttria stabilized zirconia (YSZ) and are interested in the development of YSZ coating with a specific microstructure and defect architecture. EBPVD produces coatings with a columnar microstructure and defects that, generally, are normal to metal –ceramic interface whereas the plasma-sprayed coatings exhibit a layered and porous microstructure with defects generally parallel to the metal-ceramic interface. Because of these defects, the plasma sprayed coatings exhibit a lower thermal conductivity than the EBPVD deposits but also an inferior strain tolerance [1,2].

The unique properties of coatings produced by plasma spraying result from the rapid solidification of individual molten (or partially molten) particles impinging on the substrate. At impact, the liquid spreads outward from the point of impact and forms a "splat". The process of splat formation depends on the parameters of both the impacting droplets (velocity, size, molten state, chemistry, angle of impact) and substrate surface (topography, temperature and chemistry). The dynamics of droplet deformation on the substrate determines the geometry of the resulting splat, porosity formation and quality of contact between the splat and underlying layer, thereby controlling local cooling rate and overall heat. The efficient heat extraction by the substrate results in a very high cooling rate (10^6-10^8K/s), which conditions the crystalline microstructure of the splat.
The arresting of liquid material flow on the surface results from the conversion of particle kinetic energy into work of viscous deformation and surface energy. However, solidification constraint (when the solidification front is advancing from the substrate surface fast enough to interact with the liquid during spreading) and mechanical constraint (due to the roughness of the substrate surface) can interfere with the flattening process and bring about liquid splashing..

Experimental observations and numerical simulations [3, 4] have shown that, on smooth substrates, when solidification that starts in areas where the lamella is thin enough, occurs before flattening is completed, it interferes with the spreading of the liquid material and modifies its contact with the substrate at the periphery of the solidified zone. The thin liquid material that flows outside of the solidified area is instable and may be subjected to break-up. When solidification starts after the flattening is completed, the liquid sheet formed upon flattening, and surrounded by gas, is inherently instable and disintegrates with formation of holes or waves – like disturbances along its edges (depending on the liquid viscosity).

Over the last two decades, a large body of experimental and numerical works has been carried out to better understand splat and coatings formation under plasma spray conditions. However, very few deal with impacts on rough surfaces [5]. In addition, because of the complexity of the intricate phenomena, the mechanisms are still not completely understood.
This study deals with numerical simulations of the impact of Yttria Stabilized Zirconia (YSZ) particles plasma sprayed onto smooth and rough substrates. On smooth substrates, attention is turned to the effect of the thermal properties of the substrate on splat cooling while, on rough substrates the effect of surface topography on splat formation must be also considered.

NUMERICAL MODEL

The model considers a molten ceramic particle which diameter is d_0, temperature (supposed to be uniform) T_d and velocity V_0. The droplet impacts orthogonally onto a flat smooth or rough surface (see Figure 1).

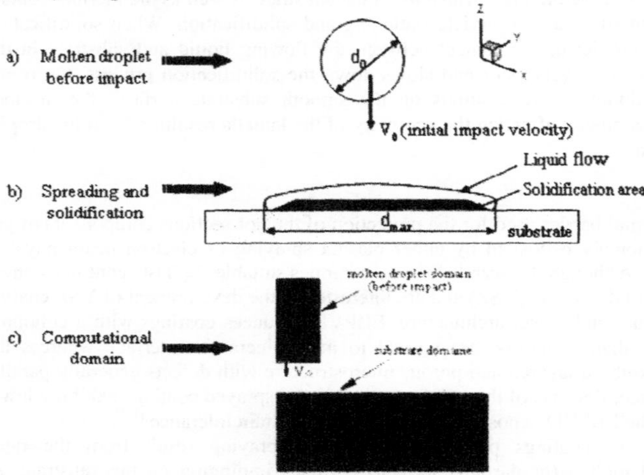

Figure 1 : Schematic description of the a) impacting molten droplet b) onto a surface substrate and c) the computational domain for both droplet and substrate.

The model [6,7] is based on the following assumptions: the fluid is Newtonian; the flow is incompressible and laminar (Reynolds numbers are small or moderate). The boundary conditions are characterized by the contact angle, the thermal contact resistance and the thermal exchange between substrate and flattening particle with the surrounding atmosphere. For instance, the quality of contact between the spreading material and the substrate is represented by a thermal contact resistance, R_{th}, which can be either constant or time dependent. In this study, a good contact is characterized by $R_{th}= 10^{-7}$ $m^2.K .W^{-1}$ and a poor one by $R_{th}= 10^{-5}$ $m^2.K .W^{-1}$. The drop wettability is characterized by the contact angle; in addition, phase change is supposed to occur at equilibrium temperature and heat exchange to occur essentially by conduction to the substrate (adiabatic boundary condition is applied at the free surface). The gaseous phase (air) is not considered in model equations, and the thermophysical properties of the droplet and substrate are temperature dependent.

Governing equations

The governing equations of the model are the conservation equations of mass, momentum, and energy that are written in 3-D cartesian coordinates and solved by using a commercial CFD code developed by Simulent [8]. The code uses a numerical difference discretization method and an interface tracking technique that utilizes a volume tracking method (volume of fluid) to follow the deformation of the free surface through the computational cells. A fraction of fluid is defined for each gird element (f = 1 if the cell is filled by the fluid, f = 0 if it is empty, and 0 < f < 1 for a cell containing a free surface). Both, the free surface and solidification interface are implicitly tracked using Eulerian formulation with a fixed grid. More details about this model can be found in Bussmann et al.[6].

Numerical procedure

To minimize the computational time, the center of the impacting droplet is located on the z-axis and calculations are performed for a quarter of droplet (x-z and y-z planes are the symmetric planes, see figure 1). The grid size (number of cells per droplet radius) and the time step were determined from preliminary calculations in order to ensure both convergence of calculations and acceptable calculation time. Under the conditions of the study, the CPR (cells per initial droplet radius) was fixed to 25.

SIMULATION INPUT PARAMETERS

The input parameters of the numerical simulations are the size, velocity and temperature of the impacting droplets, material and temperature of the substrate. Indeed, these have been experimentally found to be those controlling particle flattening and solidification [3]. In addition, two other parameters are fixed: the thermal contact resistance and the contact angle between the drop and surface. It should be noted that, in the conditions of the study, the liquid material velocity is much higher than the variation of the contact angle which, thus, does not affects significantly the flattening particle in the early stage of spreading [8].

The initial conditions used in this study are summarized in Table 1. The thermophysical data for YSZ were drawn from references [9-14]. The boundaries conditions are presented in table 2, they correspond to the conventional spray conditions.

In order to study the effect of the thermal properties of the substrate on droplet flattening and solidification processes, three substrate materials with very different thermal properties were used: copper, stainless steel and zirconia (Table 3).

Table 1: Initial parameters for numerical simulation and characteristic dimensionless numbers.

Droplet diameter (μm)	Droplet temperature (°C)	Droplet velocity ($m.s^{-1}$)	Reynolds number	Weber number
40	2800	150	1022	11666

Table 2: Main boundary conditions used in this study.

Contact Angle (°)	Interface resistance ($m^2.K.W^{-1}$)	Substrate temperature (°C)
90	10^{-5} - 10^{-7}	200

Table 3: Thermophysical properties of selected substrates at 25°C

	Properties	ρ (kg. m^{-3})	c_p (J. $kg^{-1}.K^{-1}$)	k (W. m^{-1}. K^{-1})	α (m^2. s^{-1}) x 10^{-6}
Substrate	Copper	8933	385	401	117
	Stainless steel	8055	480	15.1	2.5
	Zirconia	6570	683	2	0.44

RESULTS AND DISCUSSIONS

Impact on smooth substrates

The aim of this part is to investigate the effect of substrate material and in particular of thermal diffusivity on the flattening and solidification.

It has been shown that the substrate preheating temperature plays an important role on final splat shape by inhibiting splashing, thus bringing about a disk-shaped splat. Several explanations have been proposed to explain the existence of T_t (i) modification of substrate surface wettability, (ii) desorption of adsorbates and condensates and (iii) solidification mechanisms. Bianchi et al. [15] showed that preheating the substrate over the so-called transition temperature (T_t) improves the heat transfer at the interface splat/substrate by decreasing the thermal contact resistance, Cedelle et al. [16] also pointed out that on preheating the stainless steel substrate over T_t improves the wettability between the flattening droplet and substrate because of a change in the surface topography. In addition, Tanaka et al. [17] explained that increasing the thermal conductivity of the substrate increases the transition temperature.

Actually, the transition temperature controls the value of R_{th} which in the following has been set either to 10^{-5} or 10^{-7} m².K .W^{-1}. To study the effect of the substrate material it has been chosen to keep the substrate at 200 °C that is at about the transition temperature of stainless steel, the YSZ droplets impacting at 2800 °C (85 °C over the melting point). The final shapes of the resulting splats are presented in Fig. 2 for a good thermal contact ($R_{th} = 10^{-7}$ m².K .W^{-1}) Splats obtained on stainless steel and copper exhibit shapes close to disk except at their periphery. For stainless steel substrates, the predicted splat shape, size and flattening degree (ratio of the splat diameter to the droplet original diameter) ,are in good agreement with the experimental observations of Bianchi et al [15] and Cedelle et al. [16] In the case of zirconia substrate, part of the liquid ring detached from the central part of the splat during flattening because of much slower cooling rate.

The substrate temperature at the centre of impact (Figure 3) shows a very significant difference in temperature for the three substrate materials. In the early stage of the droplet impact, the substrate temperature rises rapidly from its initial temperature (200°C) up to a maximum value and decreases as the droplet spreads out. With copper, according to the efficient transient heat withdrawal, the substrate temperature reaches 400 °C at the maximum, while with the zirconia substrate, which diffusivity is 265 times lower, it reaches about 2000 °C and reaches an intermediate temperature for stainless steel substrate.

Of course the fast withdrawal of the enthalpy contained in the droplet by copper substrates limits drastically the development of thermal gradient within the splat, as shown in Fig. 4 a at 3.75 µs after impact. On the contrary the thermal gradient within the splat at the same time is easily noticeable on zirconia substrate, as shown in Fig. 4 c.

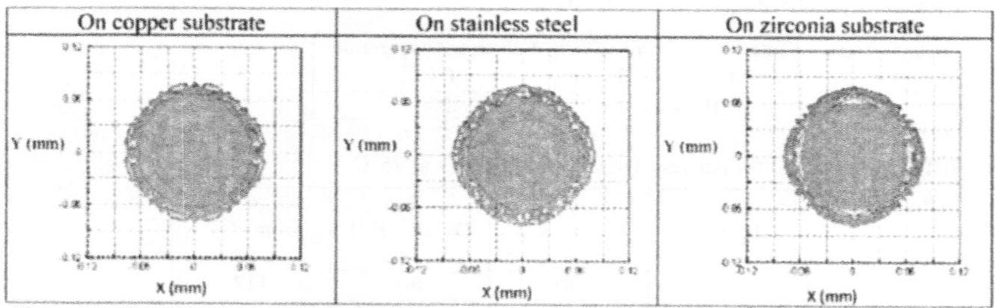

Figure 2 : Top view of predicted YSZ splat shapes after flattening (t = 10 µs) onto three different substrates a) copper, b) stainless steel and c) zirconia. The input conditions of calculations are given in Table 1. T_{sub} = 200°C and $R_{th} = 10^{-7}$ m².K .W^{-1}.

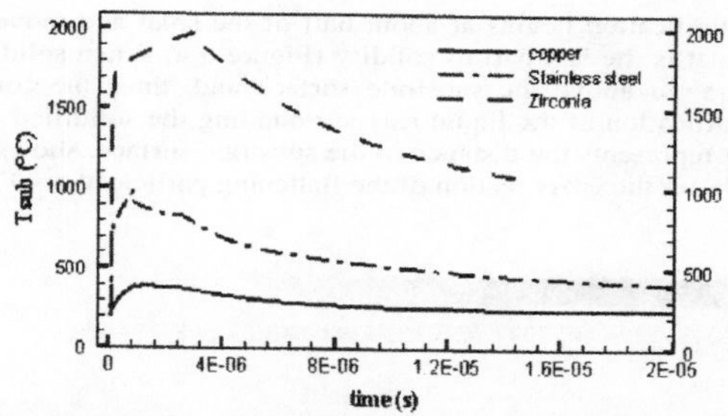

Figure 3 : Time-evolution of the substrate temperature at the centre of impact calculated for the three different substrates .Thermal contact resistance $=10^{-7}$ m².K .W^{-1}.

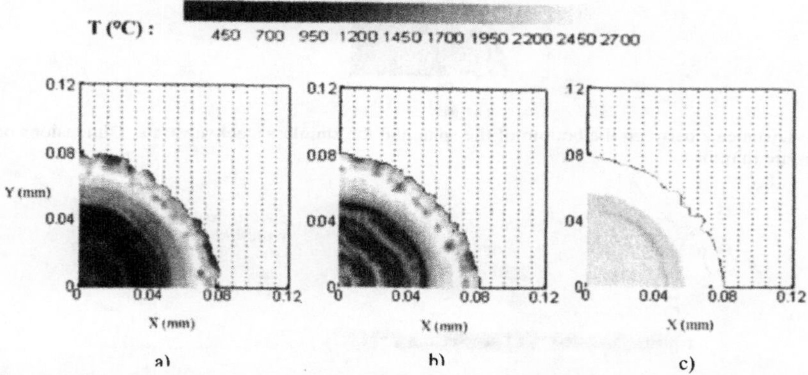

Figure 4: Temperature gradient inside YSZ splats 3.75 μs after impact onto different substrates a) copper, b) stainless steel and c) zirconia. The input conditions of calculations are given in Table 1. $R_{th} = 10^{-7}$ m².K .W^{-1}.

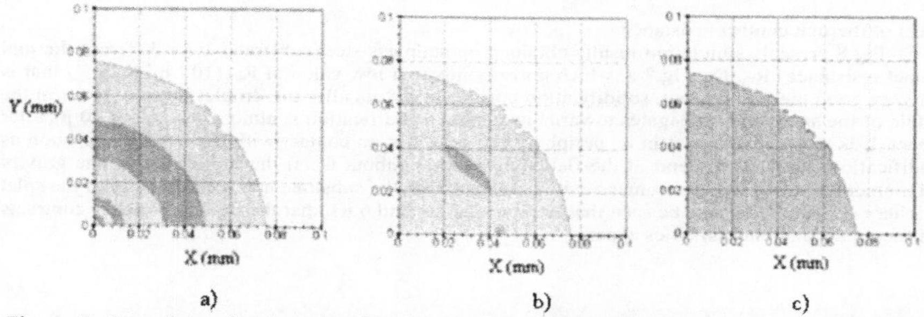

a) b) c)

Figure 5: Top view of splats showing the onset of solidification (red colour characterizes the solidified part while grey colour corresponds to the liquid material) at t = 0.875 μs inside the spreading droplet onto different substrates: a) copper, b) stainless steel and c) zirconia. The input conditions of calculations are given in Table 1. T_{sub} = 200 °C and $R_{th} = 10^{-7}$ m².K .W^{-1}.

On copper substrate, solidification begins at about half of the splat and moves towards the centre, while the edge of the splat is the last part to solidify (Figure 5 a).When solidification starts the flowing liquid is slightly raised above the substrate surface and, thus, the contact with the substrate is no more good; solidification of the liquid ring surrounding the solidified central part is, then, delayed. Fig. 6 in which z represents the distance to the substrate surface, shows the raising of the bottom of the while Fig. 7 shows the cross section of the flattening particle at t = 3 μs.

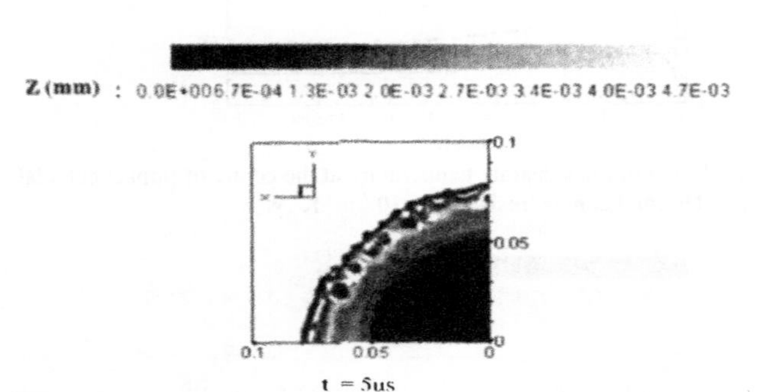

Figure 6: Distance z between the bottom of the splat and the stainless steel substrate. Dimensions on both axes are in mm.

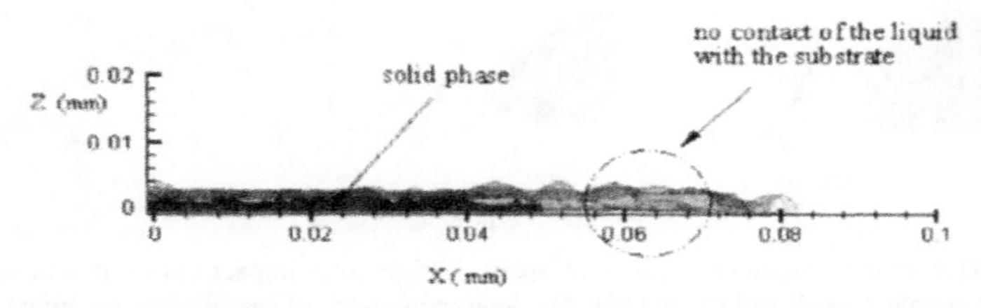

Figure 7: Splat solidification (red colour characterizes the solidified part while grey colour corresponds to the liquid material) at t = 3 μs on stainless steel substrate.

Effect of thermal contact resistance

Fig.8 presents simulation results obtained on stainless steel substrates with different thermal contact resistances R_{th}. On Fig.8 a, which corresponds to a low value of R_{th} (10^{-7} $m^2.K.W^{-1}$) that is to a very good thermal contact, solidification appears at 0.8 μs after the droplet impact, about in the middle of the splat, and propagates towards its centre. Solidification is almost completed 10 μs after impact. It is worth noting that in its periphery the splat has no contact with the substrate as soon as solidification starts. At the end of the flattening process (about 6 μs) the liquid, due to the gravity and momentum close to zero, comes again in contact with the substrate and solidification in the splat periphery occurs. It can also be seen that, at times 2.2 μs and 6 μs, that the liquid periphery contracts and fragments and internal holes appear.

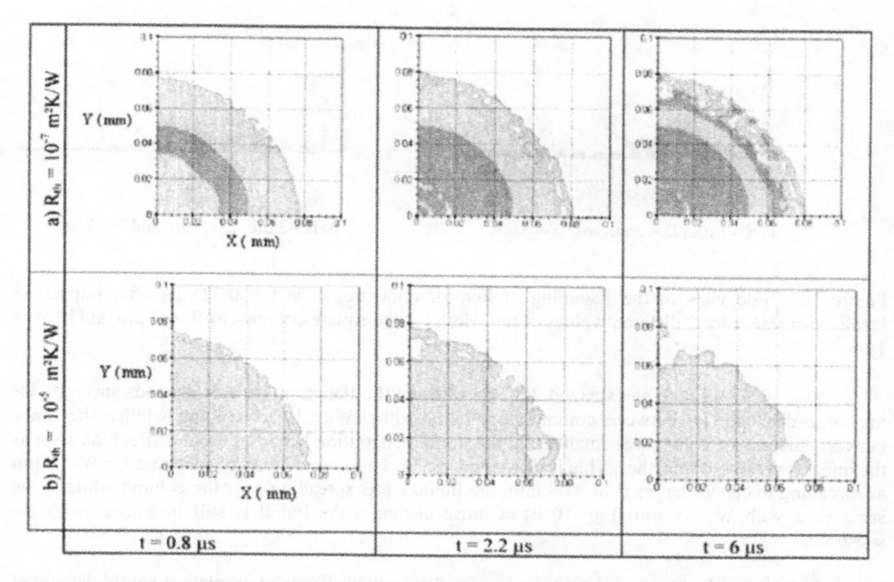

Figure 8: Top view of the splat during liquid spreading (red colour characterizes the solidified part while grey colour corresponds to the liquid material) onto a stainless substrate for a constant thermal resistance a) $R_{th} = 10^{-7} m^2K/W$ and b) $R_{th} = 10^{-5} m^2K/W$. Dimensions on both axes are in mm.

In Fig. 8 b with $R_{th} = 10^{-5}$ $m^2.K.W^{-1}$ corresponding to a poor thermal contact no, solidification occurs before 6 µs and the retraction – fragmentation of the liquid starts earlier with part of the liquid separating completely from the splat.

Impact on rough substrate

The adhesive bond strength of plasma sprayed coatings, mainly due to mechanical anchoring of the solidifying droplets, depends to a large extent on the roughness of the substrate [18]. Conventionally, the substrate surface is grit blasted to improve the mechanical anchorage, the surface roughness being usually evaluated by the average surface roughness Ra. To study the effect of substrate surface texture on the dynamic and solidification behaviour of the impacting droplet, an idealized rough surface has been considered. It is made by square columns characterized by their height, distance and width as shown in Fig. 9.

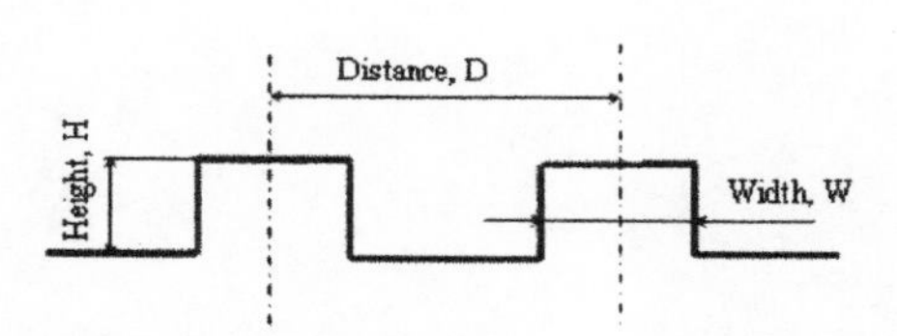

Figure 9: Idealized rough surface of substrate characterized by the three parameters: height of the square column: H, distance between column axes: D and width of the square column: W

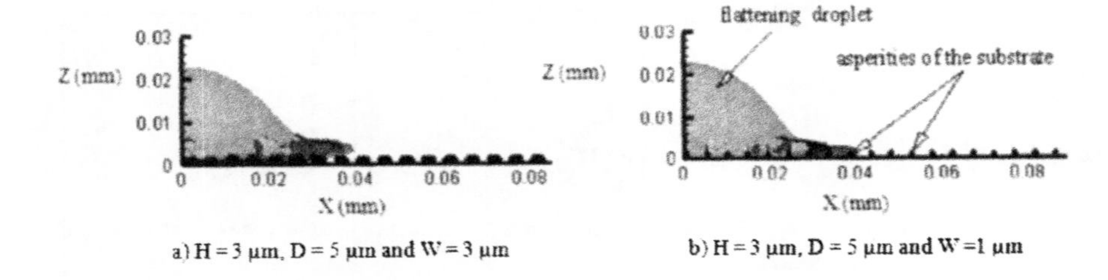

a) H = 3 µm, D = 5 µm and W = 3 µm

b) H = 3 µm, D = 5 µm and W = 1 µm

Figure 10: Side view of the flattening molten zirconia droplet at t = 0.225 µs after impact on idealized surfaces for 2 different values of the widths of the square columns a) W = 3 µm and b) W = 1 µm.

Fig. 10 shows that, as expected, the spreading of the flattening particle depends strongly, for the same distance (D) between columns, on their widths. With larger column widths, the space between successive columns is smaller and the liquid penetration becomes more difficult as soon as the impact pressure diminishes. This is illustrated in Fig. 10 a where it can be seen that for W = 3 µm at spreading distances larger than 0.03 mm, the liquid starts spreading over the column while at the same time with W = 1 µm (Fig. 10 b) at those distances the liquid is still in contact with the substrate.

Fig. 11 a to 11 f represent the top views of the partially flattened droplets at nearly the end of the flattening process (2 µs):

- For W = 3 µm (Fig. 11 a to 11 c) the fingering phenomena is marked for D = 5 µm and the liquid spreads far away as it does not penetrate the voids between substrate asperities (see Fig. 10 a). However when increasing D from 5 to 7 µm, the liquid penetration becomes easier and the finger length shortens correlatively; solidification starts 2 µs after impact at the centre of splat for larger column width (W = 7 µm).
- For W = 1 µm the fingers are mainly generated by droplets and the corresponding "matter" is less important than in previous cases; the occurs promotes splashing.

These results point out that the solidification process of droplets flattening on rough surfaces is mainly affected by the distance between the columns as it controls the ability for the spreading liquid to fill or not the voids between asperities. Fig. 12 presents, for two idealized surfaces, the liquid pressure and radial velocities at a distance z = 0.5 µm from the substrate surface at the beginning of flattening (t = 0.225 µs).

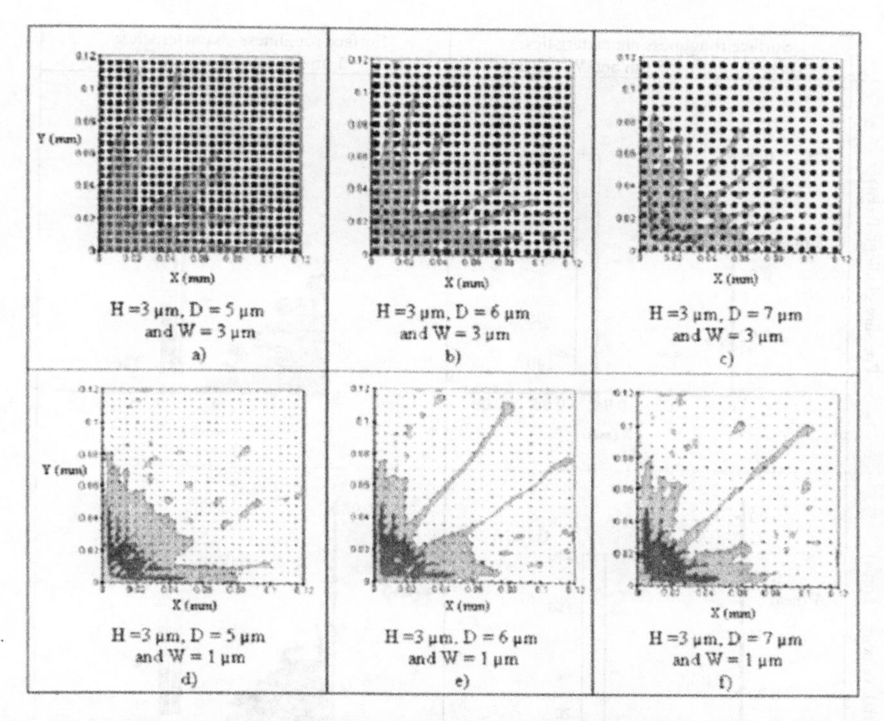

Figure 11: Comparison of spreading (grey colour) and solidification (red colour) of YSZ droplet (R_{th} = 10^{-7} m².K .W^{-1}) at the end of flattening process on substrates with different characteristics except H = 3µm for all of them. Frame a to c W = 3 µm and for d to f W = 1 µm with different values of D: a, d : D = 5 µm b, e: D = 6 µm. c, f : D = 7 µm. t = 2 µs after impact.

It can be seen in Fig. 12 a corresponding to W = 3 µm. that only very few areas exhibit pressures between 140 MPa and 1800 MP which means that area the between columns is almost free of liquid. On the contrary, it can be seen in Fig. 12 b that the whole splat area corresponding to W = 1 µm is filled with liquid with a mean pressure of about 420 MP.

The flow velocity at z = 0.5 µm. (Fig 12 c), confirms that almost no liquid is present between columns when the width of columns is 3µm and moves very slowly while when the distance between the columns is larger (Fig. 12 d) the liquid fills the inter-column space and moves radially at velocities over 200 m.s^{-1}.

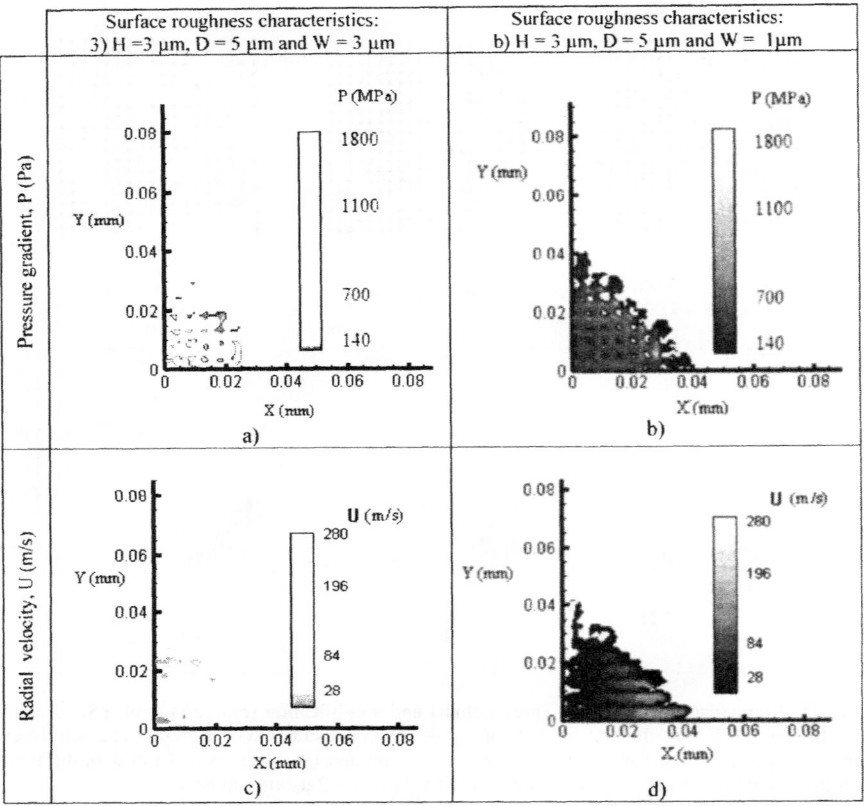

Figure 12: Liquid pressure and velocity field in a cross section of flattening droplet at z = 0.5 µm, 0.225 µs after impact on idealized surface a) W = 3 µm and b) W = 1 µm (H = 3 µm. D = 5 µm).

CONCLUSION

The flattening and solidification of fully molten YSZ droplets (2800°C) 40 µm in diameter impacting on smooth and rough substrates at 150 m/s have been numerically studied. Three substrates with different thermal diffusivities (maximum ratio 267) have been considered as well as different thermal contact resistance (10^{-7} and 10^{-5} $m^2.K.W^{-1}$). The substrate roughness has been idealized with square columns with different widths for the same heights and distances between columns. Results have emphasized the importance of substrate diffusivity on the temperature gradient within the formed splat. The solidification depends strongly upon substrate diffusivity and thermal contact resistance between splat and substrate. As soon as solidification has started (generally in the flattening splat periphery) the liquid flow, beyond the solidified area, has no contact with the substrate. The contact is re-establish further when the momentum of the liquid tends to zero. When the thermal contact resistance is higher (10^{-5} $m^2.K.W^{-1}$) nearly no solidification occurs before spreading is completed and the liquid contacts and breaks up. On rough substrates splats are extensively fingered and the fingering phenomenon is strongly linked to the distance between the square columns representing the substrate surface roughness.

NOMENCLATURE

d_p : initial droplet diameter (μm)
T_m : melting temperature of the droplet (°C)
T_{sub} : substrate temperature (°C)
T_t : transition temperature (°C)
T_d : initial droplet temperature (°C)
V_0 : initial droplet impact velocity ($m.s^{-1}$)
v: kinematic viscosity of liquid material ($m^2. s^{-1}$)
R_{th} : thermal contact resistance lamella and substrate($m^2.K .W^{-1}$)
We: Weber number = $\rho V_0^2 d_p/\sigma$
Re: Reynolds number = $V_0 d_p/v$
ρ: specific mass ($kg. m^{-3}$)
σ: molten particle surface tension ($J. m^{-2}$)
c_p : specific heat ($J.kg^{-1}.K^{-1}$)
k: thermal conductivity ($W.m^{-1}.K^{-1}$)
α: thermal diffusivity ($m^2.s^{-1}$)
U: radial velocity in x-direction: component of velocity field ($m. s^{-1}$)

REFERENCES

[1]D.D. Hass, P.A. Parrish, H.N.G. Wadley, Electron beam directed vapor deposition of thermal barrier coatings, Journal of Vacuum Science and Technology, vol A 16(6), 3396-01(1998).
[2]A. Feuerstein, J. Knapp, T. Taylor, A. Ashary. A. Bolcavage, N. Hitchman, Thermal Barrier Coating Systems for Gas Turbine Engines by Thermal Spray and EBPVD - A Technical and Economic Comparison, Proceedings of the 2006 International Thermal Spray Conference, May 15-18, 2006, Seattle, Washington, USA, Copyright © 2006 ASM International®
[3]P. Fauchais, M. Fukumoto, A. Vardelle, M. Vardelle, J. Therm. Spray Technol. 13, 337 (2004).
[4]Javad Mostaghimi and Sanjeev Chandra, Splat formation in plasma-spray coating process *Pure Appl. Chem.*, Vol. 74, No. 3, 441–445 (2002).
[5] H.B. Parizi, L. Rosenzweig, J. Mostaghimi, Numerical Simulation of Droplet Impact on Patterned Surfaces, J. of Thermal Spray Technology, vol 16 (5-6), 603-1020 (2007).
[6]Bussmann, M., Mostaghimi, J. and Chandra, S., On a three-dimensional volume tracking model of droplet impact, Physics of Fluids, 7, 1406-17 (1999).
[7] http://www.simulent.com
[8] N. Ferguen, A. Vardelle and P. Fauchais, Numerical study of splat solidification in a plasma spray deposition process, Proceedings of the 5th Decennial International Conference on Solidification Processing, Sheffield (UK), 632-635 (2007).
[9]K. Shinoda, T. Koseki, and T. Yoshida, Influence of impact parameters of zirconia droplets on splat formationand morphology in plasma spraying, JOURNAL OF APPLIED PHYSICS **100**, 074903 (2006).
[10]http://www.nist.gov/
[11]M. Lihrmann and J.S. Haggerty, "Surface Tensions of Alumina-Containing Liquids," J. Am. Ceram. Soc., 68 (2), 81-85 (1985).
[12]G.H. Geiger and D.R. Poirier: Transport Phenomena in Metallurgy, Addison-Wesley Publishing Company, Reading, MA (1973).
[13]Y.S. Touloukian: Thermophysical Properties of Matter, IFI/Plenum, New York, NY, 79 (1970).
[14]Y. Kawai and Y. Shiraishi: Handbook of Physico-Chemical Properties at High Temperatures, Iron and Steel Institute of Japan, Tokyo,(1988).
[15] Bianchi, L. and Leger, A. C. and Vardelle, M. and Vardelle, A. and Fauchais, P., Splat formation and cooling of plasmasprayed zirconia, Thin Solid Films, 35-47, 305 (1997).

[16]Cedelle, J. and Vardelle, M. and Pateyron, B. and Fauchais, P. Fauchais, Experimental Investigation of the Splashing Processes at Impact in Plasma Sprayed Coating Formation, Proceedings of the 2004 International Thermal Spray Conference, 10-12 Mai, Osaka (Japon), ASM International,(Pub.) (2004).
[17] Y. Tanaka a, M. Fukumoto, Investigation of dominating factors on flattening behavior of plasma sprayed ceramic particles, Surface and Coatings Technology 120–121 , 124–130 (1999).
[18] M. Mellali, A. Grimaud, A.C. Leger, P. Fauchais and J. Lu, Alumina grit blasting parameters for surface preparation in the plasma spraying operation, J. of Thermal spray Technology, 6 (2), 217-227 (1997).

Author Index

Printed and bound by CPI Group (UK) Ltd, Croydon, CR0 4YY

16/04/2025

14658453-0005